Petroleum Engineering

Sustainable Geoenergy Engineering and Technology

The objective of this series is to establish a link between traditional hydrocarbon science and engineering and the development of sustainable engineering to achieve net-zero and sustainability goals in energy production. The series aims to promote and expedite the dissemination of new research results, innovations, and technology in the field of exploration, production, and management of oil and gas, aiding the oil and gas industry in navigating through the energy transition and driving it towards the net-zero target. Additionally, the series covers developing areas related to the geological subsurface, such as geothermal energy and underground storage. It includes monographs, lecture notes, and edited volumes, with a subject focus on upstream processes. Coverage extends to all theoretical and applied aspects of the field, encompassing both traditional and modern topics, including but not limited to:

Exploration Formation evaluation (well logging) Drilling Economics Reservoir simulation Reservoir engineering Well engineering Artificial lift systems Facilities engineering Sustainable energy exploration and production Carbon capture, utilization, and storage (CCUS) Geothermal energy development Underground Hydrogen Storage (UHS) Information technology (IT)

Contributions to the series can be made by submitting a proposal to the responsible Springer contact, Anthony Doyle anthony.doyle@springer.com

Chaohua Guo · Zhao Yang

Shale Gas Production: Concept, Models, and Techniques

A Comprehensive Study of Fluid Transport in Shale Gas Reservoirs

Springer

Chaohua Guo
Faculty of Energy Resources
China University of Geosciences (Wuhan)
Wuhan, Hubei, China

Zhao Yang
Department of Petroleum Engineering
Northeast Petroleum University
Daqing, Heilongjiang, China

ISSN 2366-2646 ISSN 2366-2654 (electronic)
Petroleum Engineering
ISBN 978-3-031-84551-2 ISBN 978-3-031-84552-9 (eBook)
https://doi.org/10.1007/978-3-031-84552-9

This Springer imprint is published by the registered company Springer Nature Switzerland AG
The registered company address is: Gewerbestrasse 11, 6330 Cham, Switzerland

Preface

This textbook is written for advanced undergraduate and graduate students in petroleum engineering, as well as industry practitioners seeking to deepen their understanding of shale gas production. The content has evolved from years of teaching and research, incorporating feedback from students and colleagues to create a cohesive and accessible learning resource.

The structure of the book facilitates a progressive learning experience. Chapter 1 establishes a foundational understanding of shale gas concepts, while Chap. 2 examines pore structure characteristics that significantly influence reservoir behavior. Subsequent chapters progressively introduce gas transport mechanisms, culminating in comprehensive models for both single and multi-phase gas transport. Chapter 3 explores gas transport mechanisms in the nanopores of shale gas reservoirs, detailing how molecular behavior impacts flow dynamics at the nanoscale. This chapter provides insights into the unique challenges and opportunities presented by the small-scale structure of shale formations. Chapter 4 introduces the pure gas transport model, focusing on the theoretical frameworks that govern gas movement in nanopores. Readers will gain an understanding of the mathematical representations and assumptions necessary for modeling gas flow in these confined spaces. In Chap. 5, we summarized the two-component gas transport mechanisms in shale gas nanopores. This chapter provides a comprehensive summary, including the general flow mechanisms, adsorption models, and existing models for two-component gas transport. In Chap. 6, we delve into the two-component gas transport model, addressing how the interaction between different gas phases affects overall transport processes. This chapter highlights the complexities of gas mixtures in shale reservoirs, which are crucial for optimizing extraction strategies. Chapter 7 examines the two-phase gas and water transport model, illustrating the interplay between gas and water in shale gas production. This chapter emphasizes the importance of understanding fluid interactions to improve production efficiency and manage reservoir behavior effectively. Chapter 8 focuses on practical applications, discussing multi-stage horizontal well hydraulic fracturing techniques and models, highlighting their critical role in enhancing production efficiency. Chapter 9 is the conclusion section.

We believe that this book will not only serve as an educational tool but also inspire further exploration and innovation in shale gas production techniques.

Wuhan, China Chaohua Guo
Daqing, China Zhao Yang

Introduction

The rise of shale gas as a critical energy resource has transformed the global energy landscape. With advancements in extraction technologies and a deeper understanding of reservoir dynamics, shale gas production has become a vital component of energy strategies worldwide. This book aims to equip readers with foundational knowledge and technical expertise necessary for engaging with the complexities of shale gas production.

In this text, we explore the multifaceted nature of shale gas reservoirs, beginning with essential concepts and pore structure characteristics that define these unique geological formations. We delve into the intricate gas transport mechanisms within the nanopores of shale, presenting models that address pure gas transport, two-component gas interactions, and the interplay between gas and water phases. Each chapter is designed to build upon the last, leading to a comprehensive understanding of multi-stage horizontal well hydraulic fracturing techniques and their applications in optimizing shale gas extraction.

The journey through this book will not only enhance theoretical knowledge but also provide practical insights into the modeling techniques that underpin successful shale gas production. It is our hope that this resource will serve as a cornerstone for students, researchers, and industry professionals alike, fostering a deeper appreciation for the science and technology driving this dynamic field.

Contents

About the Authors

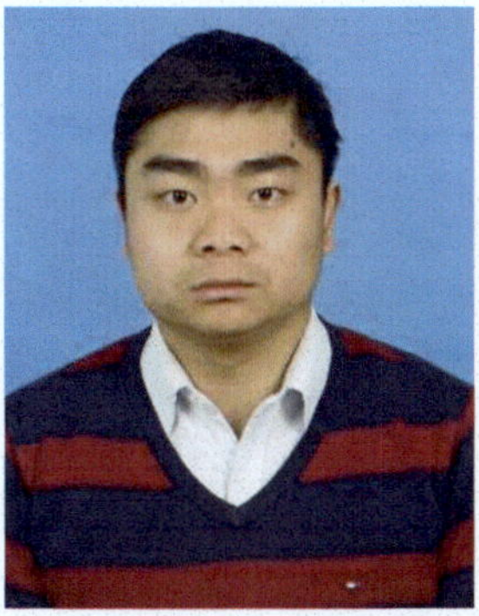

Chaohua Guo is Associate Professor in China University of Geosciences (Wuhan). Dr. Guo has more than 10 years experiences of research and teaching in unconventional shale gas production. He has led and undertaken the National Natural Science Foundation of China and more than 10 projects from oil companies. As the first/corresponding author, he has published more than 40 SCI/EI papers in journals such as SPE Journal, FUEL, Journal of Petroleum Science and Engineering, Petroleum Science & Engineering, Physics and Chemistry of the Earth. He also served as the reviewer for more than 20 journal papers, such as Interpreter; Industrial & Engineering Chemistry Research; Journal of Computational and Applied Mathematics; Chemical Engineering Journal; Open Geosciences; Journal of Petroleum Science and Engineering.

Zhao Yang is Professor in Northeast Petroleum University. He has engaged in theoretical and technical research in oil and gas field development engineering (reservoir and residual oil description, low-permeability oilfield exploitation, and improving oil and gas recovery). He has participated in the completion of two national science and technology major projects, two natural science foundation projects, and five provincial, ministerial, and oilfield research projects. He has published over 20 academic papers and over 10 authorized invention patents and utility model invention patents. He served as the editor-in-chief of one textbook titled 'Oil Layer Physics', which won the 2017 Excellent Textbook Award from the university. He also published over 20 papers in domestic and international journals, including five papers indexed by EI and ISTP.

Chapter 1
Basic Concept with Shale Gas

1.1 Shale Gas

Shale gas is the natural gas which is produced from those tight shale formations. Shale gas is becoming an increasingly important factor to secure energy over recent years for the considerable volume of natural gas stored and is gradually becoming a key component in the world's energy supply. Shale strata or shale gas reservoirs are typical unconventional reservoirs with critically low transport capability in matrix and numerous "natural" fractures. Unconventional reservoirs are those geological formations which contain fossil energy but are uneconomical to produce (with recovery about 2%) currently when conventional recovery methods are applied, such as shale oil/gas reservoirs, tight oil/gas reservoirs, and coalbed methane reservoirs. Shales are fine-grained (with average grain size below 0.0025 in) sedimentary rocks with ultra-low permeability (in the order of nanodarcy). Typical shale gas reservoirs exhibit a net thickness of 50–600 ft, porosity of 2–8%, total organic carbon of 1–14% and depth of 1000–13,000 ft. They are organic-rich formations and are both the source rock and the reservoir. Natural gas have been stored in shale gas reservoirs with three forms: (a) in the micro fractures and nano pores, they are stored as free gas; (b) in the kerogen, they are stored as dissolved gas; (c) in the surface of the bedrock, they are stored as adsorption gas and about 20–85% of gas in SGRs are stored in this kind of form.

1.1.1 Importance of Shale Gas Reservoirs

Natural gas, coal, and oil contribute 85% to the whole nation's energy structure, and natural gas contributes 22% of the total. Natural gas is the cleanest fossil fuel compared with oil and coal. Most of the technically recoverable natural gas in North America is present in unconventional reservoirs such as tight sands, shale,

C. Guo and Z. Yang, *Shale Gas Production: Concept, Models, and Techniques*, Petroleum Engineering, https://doi.org/10.1007/978-3-031-84552-9_1

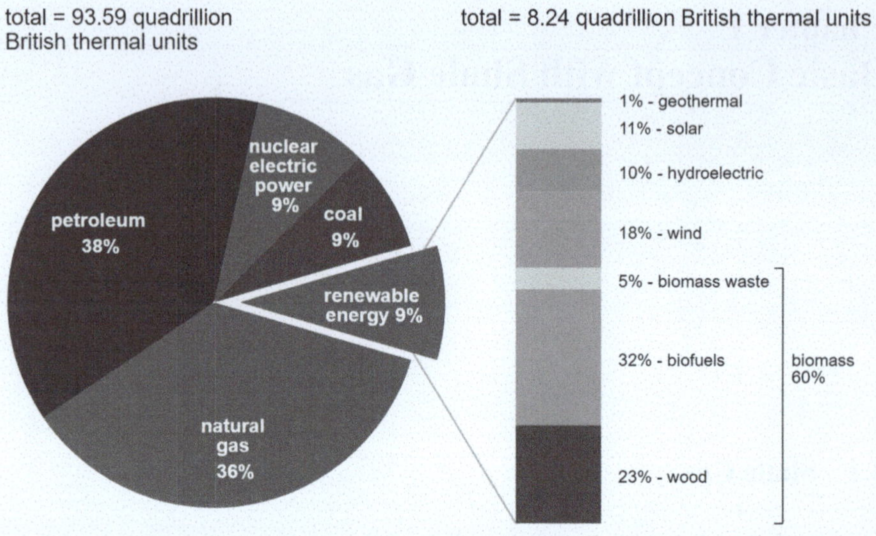

Fig. 1.1 U.S. primary energy consumption by energy source [2]

and coalbed. Natural gas plays an important role in satisfying U.S. energy demands and contribution of natural gas will continue to increase in the next twenty years, and is expected to increase to 38% of the total in 2040 (Fig. 1.1). In 2012, shale gas has contributed 31% to the total U.S. natural gas production. Also, North America has the largest shale gas reserves. And the total shale gas volumes around the world are estimated to be about 16,000 Tcf. Owing to new applications of horizontal well drilling, multi-stage hydraulic fracturing, and other technologies, shale gas has offset declines in production from conventional gas reservoirs, and contributes to major increases of U.S. natural gas production. Overall, natural gas production from unconventional reservoirs, especially shale gas reservoirs, is becoming an ever increasing portion of the U.S. main natural gas supplier, securing the bulk of the U.S. natural gas supply for the next twenty years which is shown in Fig. 1.1 (EIA 2023). And the U.S. Energy Information Administration (EIA) states that "natural gas from tight sand formations is currently the largest source of unconventional production, accounting for 30% of total U.S. production in 2030, but production from shale formations is the fastest growing source." [1]

1.1.2 Pore Size Distribution of Shale Gas Reservoirs

Understanding the pore size distribution is of great importance in evaluating the original gas in place and flow characteristics of the shale matrix. It is found that most pore throat diameters are concentrated in the range of 4–200 nm through experimental analysis on 152 cores from nine reservoirs in North America. And the pore sizes

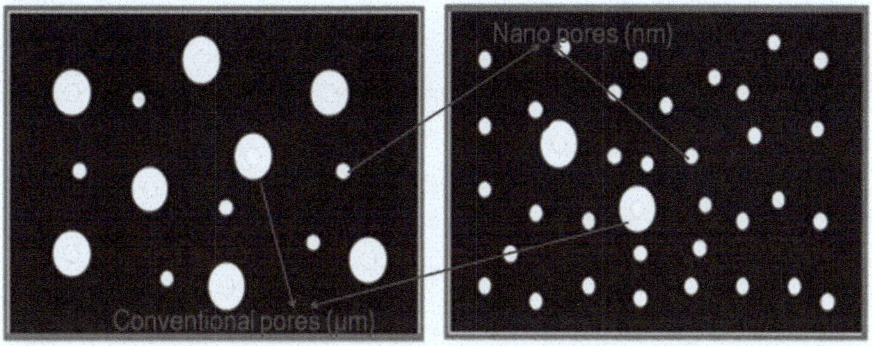

Fig. 1.2 Comparison between conventional gas reservoir (**a**) and shale gas reservoir (**b**)

are in the range of micro to meso scale, with average less than 10 nm. Figure 1.2 is the pore distribution comparison between conventional and unconventional (shale) reservoirs [3]. It can be found that in unconventional shale gas reservoirs, the number of nanopores is higher than those in conventional reservoirs. Due to this, the exposed surface area in shale gas reservoirs is larger than those in conventional reservoirs, which leads to more adsorption gas. Also, gas flow in those nano pores is different from Darcy flow. The diameter of pores in shale gas reservoirs are from a few nanometres to a few micrometres. And the morphology of the nano pores are also different. Figure 1.3 is the scanning electron images of the shale sample [4]. From the figure, we can find that for the nano pores, there are four different morphologies.

(1) Very small pores. The pore size is from 18 to 46 nm. The nanopores are nearly spherical. Total porosity is about 5.2%.
(2) Larger nanopores. The pore diameter is about 550 nm.
(3) Tubelike pore throats connecting elliptical pores. The pore size is less than 20 nm.
(4) Additional tubelike pore throats connecting elliptical pore. The pore-throat diameter is less.

Wang et al. [5] have conducted pore structure study on Barnett shale in the Fort Worth Basin, North Texas, and have classified the shale formation into four different medium: organic matrix, nonorganic matrix, natural fractures, and hydraulic fractures. Researchers also classified the matrix pores according to the scale. Two types of matrix pores are nano-scale pores and micro-scale pores. Shale gas can not only be stored as adsorption gas on the surface of organic grains, but also can be stored as free gas inside the organic grains. They found that the pore network in the organic matrix are important gas flow paths. Sondergeld et al. [1] have found that the pores within the Kerogen of shale matrix are in the range of mesopores from 5 to 23 nm using scanning electron microscope imaging (SEM). Curtis et al. [6] have detected pores about 2 nm using scanning transmission electron microscope imaging (STEM). All these studies have revealed the widely existence of nano pores in tight shale formations.

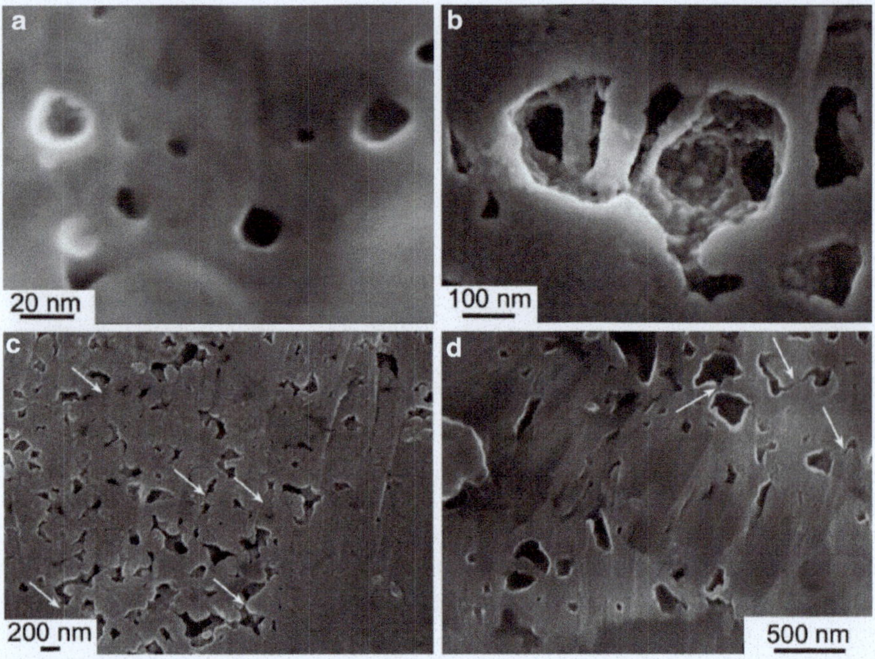

Fig. 1.3. Variations in nanopore morphology in organic matter [4]

1.1.3 Gas Storage Forms in Shale Gas Reservoirs

Understanding gas storage forms is of great importance for the understanding of the gas transport process in the reservoir. Different from conventional reservoirs where gas is only stored in the pore space as free gas, unconventional shale gas reservoirs have a variety of storage forms. There are three forms for gas to be stored in shale gas reservoirs. (a) Adsorption gas in the organic matter; (b) free gas in the organic pores, non-organic pores, and micro-fractures; (c) stored gas in the hydraulic fractures. Javadpour [7] suggested that there are also three forms (Fig. 1.4) for gas to be stored in shale gas reservoirs: (a) in the micro fractures and nano pores, they are stored as free gas; (b) in the kerogen, they are stored as dissolved gas; (c) in the surface of the bedrock, they are stored as adsorption gas. Figure 1.5 illustrates the gas distribution and geometry of shale strata from micro to macro scale. It informs us that in the fracture, only free gas exists and in the matrix which is full of kerogen, free gas and adsorption gas co-exist.

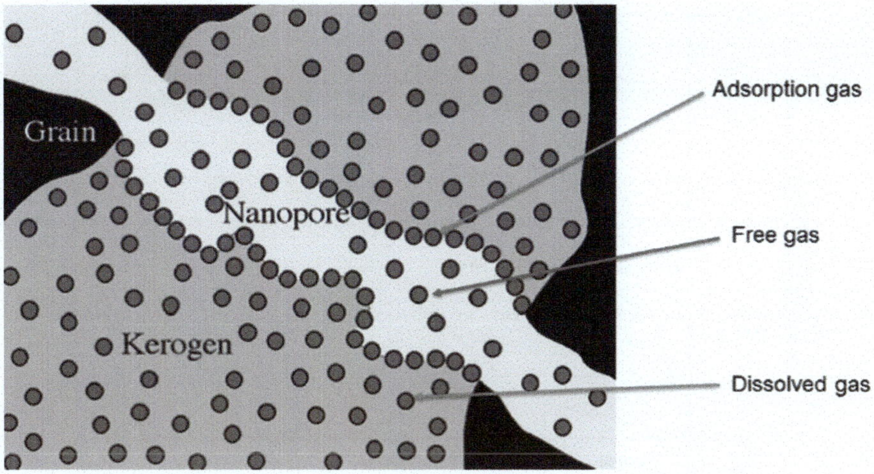

Fig. 1.4 Gas storage forms in shale gas reservoirs [3]

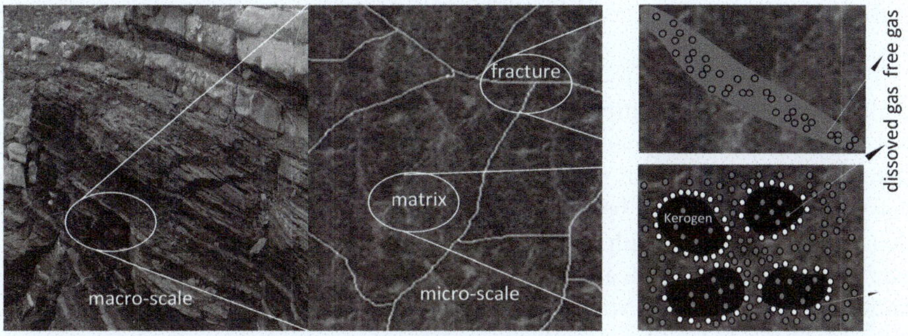

Fig. 1.5 Gas distribution in shale strata from macro-scale to micro-scale. In the fracture there exist free gas and, in the matrix, free gas and adsorption gas co-exist

1.1.4 Permeability and Porosity of Shale Gas Reservoirs

Permeability and porosity are two critical parameters to determine the flow capacity and gas content in place. As hydrocarbon formation, shales have low permeability and low porosity, which are as impermeable as concrete with permeability in the nanodarcy and porosity between 2 and 10%. Also, through the experimental analysis of 152 cores of nine reservoirs in North America, Javadpour found that the average permeability of shale bedrock is 54 nd, and approximately 90% have permeability less than 150 nd [7]. Wang et al. [5] have a made a survey on the permeability–porosity relationship for shale gas plays in North America. The result of the survey is illustrated in the Fig. 1.6.

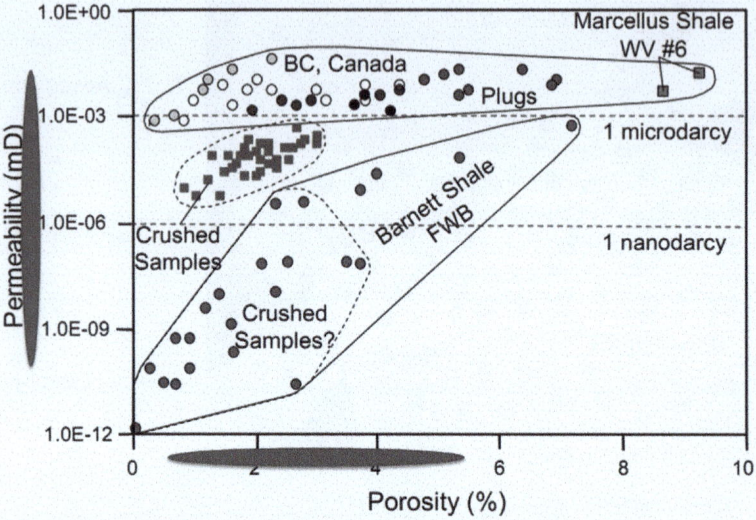

Fig. 1.6 Survey of permeability and porosity for shale gas plays in North America [5]

From Fig. 1.6, it can be found that the porosity and permeability for North America shale gas reservoirs are extremely low. The matrix permeability is from 1 nanodarcy to 1 microdarcy, and the matrix porosity is 1–5%. Porosity can determine the storage state of shale gas: free gas or adsorption gas. In those shale formations with high porosity, gas will mainly be stored as free gas. And gas will be stored as adsorption gas in formations with low porosity. In the Ohio shale and Atrim shale with average porosity of 5–6% and maximum porosity of 15%, the free gas can be up to 50% of the total pore volume. Methods used to obtain the ultra-low shale gas permeability include pulse-decay technique, crushed samples dynamic pycnometry, canister desorption test.

1.1.5 Accumulation and Production Characteristics of Shale Gas Reservoirs

Understanding of the accumulation process is important for the understanding the gas distribution and the understanding of production process is key for the construction of numerical simulation model. The formation process of the shale gas reservoirs is totally different from conventional reservoirs. In conventional reservoirs, the hydrocarbons migrate from the source rock to the reservoir rock. When hydrocarbons formed, they will migrate and stops when it finds a structural trap which is called the reservoir rock. Then, the cap rock seals the hydrocarbons to prevent its further migration. Commonly, the cap rock are shale with extremely low permeability. However, due to the ultra-low permeability of shale gas reservoirs, there is no migratory path for the produced gas after it is formed in shale gas reservoirs. The shale formation are both the source rock and the reservoir rock in itself. The formation process of shale gas reservoir is illustrated in Fig. 1.7, which can be divided into three periods: (1) The organic matter (Kerogen) gets deposited and the natural gas produced from natural fractured shale; (2) when the adsorption gas and dissolved gas become saturated, the extra gas will desorb and goes into the matrix pores; (3) with large amount of natural gas produced, the matrix pressure increase. Free gas flows into the fracture and accumulated. The production process of shale gas reservoirs is a reverse process compared with the formation process. Javadpour has studied the production process of shale gas reservoirs, which is shown in Fig. 1.8 [3]. When producing the shale gas, firstly, the free gas in the fractures and large pores will be produced, see the step (4) in Fig. 1.8. And then gas will be produced from the smaller pores, see the step (3) in Fig. 1.8. With pressure depletion, the thermodynamic equilibrium between Kerogen and gas phase in the pores will change and gas will desorb from the surface of the Kerogen, see the step (2) in Fig. 1.8. Due to the pressure difference, gas molecules will diffuse from the bulk of Kerogen to the surface, see the step (1) in Fig. 1.8.

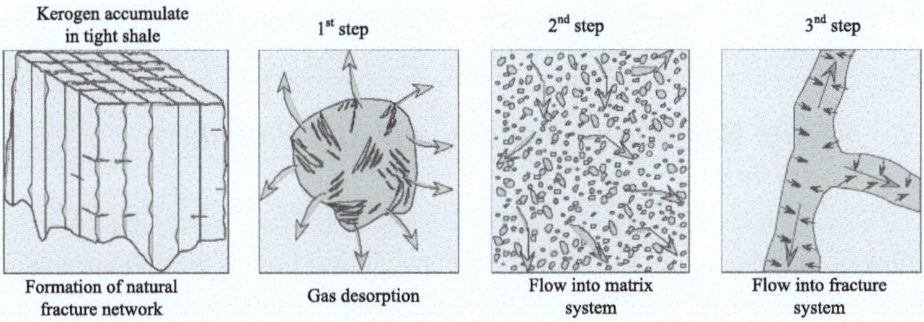

Fig. 1.7 Formation process of shale gas reservoir

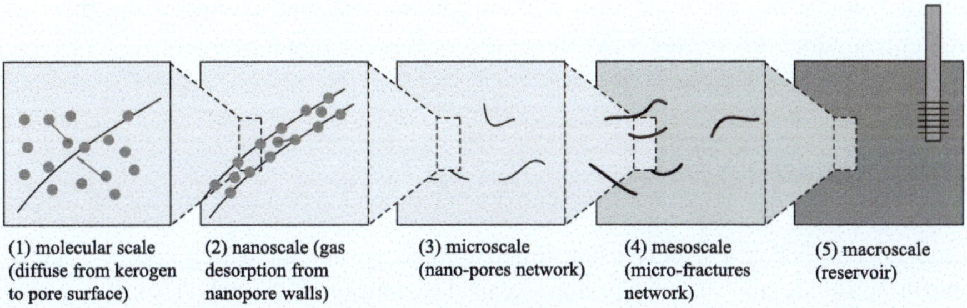

Fig. 1.8 Gas production process from macro to molecular scales. Flow to the wellbore is first initiated at the macroscale, followed by flow at increasingly finer scales

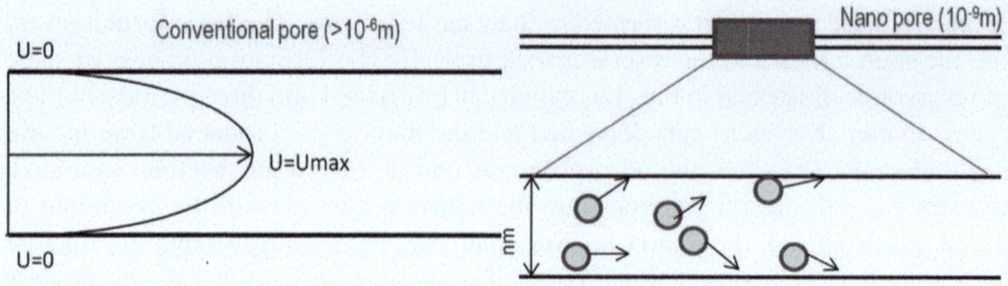

Fig. 1.9 Comparison between no-slip and slip boundary condition

1.2 Gas Flow Behavior in Nano Pores

Due to the tiny pore size, the fluid continuum theory breaks down for shale pores with pore throat diameters in the range of nano scale. Some smaller pores, such as 5–50 nm are also found which is in the same scale of the kinetic diameter of methane molecules. Except the conventional convective flow, many other flow mechanisms exist in the nano pores.

1.2.1 No-slip Boundary Condition and Slip Boundary Condition

Gas flow in the nano pores is different from conventional flow as the no-slip boundary condition will no longer be valid when the length scale of a physical system decreases. The difference between no-slip and slip boundary condition is shown in Fig. Fig. 1.9. In the no-slip boundary condition, the velocity profile in the pore is parabolic and the velocity on the wall is equal to zero. However, in the slip boundary such as in the nano pores, the gas molecules will slip on the wall and collide with other gas molecules, which means the velocity on the wall is not equal to zero.

1.2.2 Knudsen Number

Knudsen number is a widely used dimensionless parameter which can be used to determine the degree of appropriateness of the continuum model. Also, it is a tool to determine the limitations of the Navier stokes equation according to the existence of slip or no-slip boundaries. The Knudsen number is defined as the ratio of gas mean-free-path λ to the pore throat diameter d. And the gas mean free path is defined as the distance required along the straight direction until that gas molecule collisions with other molecules or solid wall occurs. The formula to calculate gas mean free path is shown in Eq. (1.1) which is provided by Jeje and Mohtadi [8].

$$K_n = \frac{\lambda}{d} \tag{1.1}$$

$$\lambda = \frac{k_B T}{\sqrt{2}\pi \delta^2 P} \tag{1.2}$$

where K_n is the Knudsen number, λ is the gas mean free path, d is the pore diameter, k_B is the boltzmann constant (1.3805×10^{-23} J/K), T is the temperature (K), P is the pressure (Pa) and δ is the collision diameter of the gas molecular. It can be found that K_n is related to pore throat diameter, gas pressure, and the temperature. It is positive related to temperature and negative related to pressure. Some scholars also proposed some other forms of equations to calculate the Knudsen number for simplification or application.

Roy et al. [9] have provided following formula to calculate the gas mean free path:

$$\lambda = \frac{16\mu}{5} \frac{1}{\rho\sqrt{2\pi R_g T}} \tag{1.3}$$

Also, Loeb [10] has proposed following formula to calculate gas mean free path considering ideal gas which is expressed as follows:

$$\lambda = \frac{\mu}{\rho}\sqrt{\frac{\pi}{2R_g T}} = \frac{\mu}{p}\sqrt{\frac{\pi R_g T}{2M}} \tag{1.4}$$

where μ is the gas bulk viscosity, ρ is the gas density, M is the gas molecular mass, R_g is the universal gas constant, T is the gas temperature, and p is the pore pressure.

A more fundamental definition of gas mean free path is presented as follows [11]:

$$\lambda = \frac{1}{\sqrt{2}\pi (N_m/V) d_m^2} \tag{1.5}$$

where N_m is the number of molecules, V is the volume occupied by the molecules and d_m is the diameter of a molecule. If the molecule cannot be described by a rigid sphere, πd_m^2 should be replaced by the total scattering cross section.

Knudsen number represents the ratio between viscous flow and diffusion. According to the Knudsen number, the gas diffusion can be classified into molecular diffusion, Knudsen diffusion, and surface diffusion. When the gas mean free path is less than the pore diameter, molecular diffusion dominates. The flow is dominated by the collision between gas molecules. When the pore diameters is greatly less than the gas mean free path, Knudsen diffusion dominates. The flow is dominated by the collision between gas molecules and the wall.

Also, Knudsen number is a measure of the gas rarefaction degree when gas flow through small pores. According to the Knudsen number, the gas flow regime can be classified into four kinds, which is illustrated in Fig. 1.10.

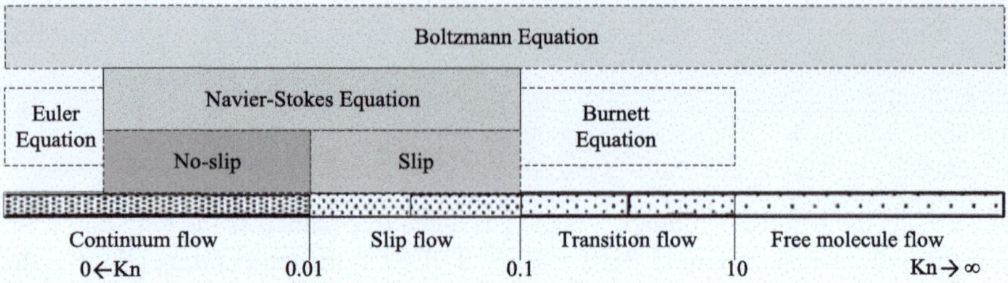

Fig. 1.10 Flow regime classification based on Knudsen number [12]

According to Knudsen number, the fluid flow regimes can be classified into four types:

(1) $K_n < 0.01$. Continuum regime. Under this Knudsen number range, the continuum assumption holds. The no-slip boundary condition is valid. And the flow is continuum flow or viscous flow. For simplification, gas transport can be characterized using conventional Darcy law.
(2) $0.01 < K_n < 0.1$. Slip flow regime under this Knudsen number range. The no-slip boundary condition is not valid any longer and the rarefaction effects become more pronounced. The flow is called slip flow which cannot be characterized using continuum approach. The ratio of molecule-wall to molecule–molecule collisions increases as Knudsen number increases. However, the latter is still dominant in the slip flow regime. Therefore, the Navier–Stokes equation is still valid for this domain. The flow can be described using Navier–Stokes equation combined with slip boundary condition. Also, the slip flow regime can be modeled using Dusty Gas Model.
(3) $0.1 < K_n < 10$. Much higher Knudsen number. The flow is called transition flow regime. The continuum assumption and Navier–stokes equation are both not applied. Molecular simulation method can be used to study this flow regime, in which the gas molecules are considered as a swarm of discrete particles.
(4) $K_n > 10$. The intermolecular collisions become negligible and the flow regime is called free molecule flow. In this Knudsen number range, the gas mean free path of the gas molecules is 10 times larger than the pore size. Boltzmann equation can be used to characterize such flow regime. However, it should be noted that this classification based on Knudsen number is empirical and the bounds of different regimes should not be exact values. Some literatures have set the limit of viscous flow region as 0.001.

Many investigators have studied micro-scale flow in shale nano pores. Civan and Vafai [13] has separated the gas flow in a single capillary flow path according into three flow regimes according to the distance away from the pore wall. The flow path has been classified into three regions: inner flow region, intermediate region, and the near wall region. In the near wall region, gas transport can be described by a constant slip velocity and can be characterized as free molecular regime. In the intermediate region, gas transport can be characterized using transition/Knudsen flow regime. In the inner region which is away from the pore wall, the flow can be characterized as continuum flow. Yucel Akkutlu and Fathi [14] developed a model to consider desorption and diffusion from Kerogen. Swami and Settari [15] have proposed a model for gas flow through one nano pore in shale gas reservoirs considering the effects of non-Darcy flow mechanism. According to the literature review, the most widely accepted flow mechanisms in gas shales are summarized as follows:

(1) Viscous flow or convective flow: This is the region where Knudsen number has fairly low values such as in the conventional reservoirs where pore diameter is in micrometers. The gas mean free path is comparably negligible with the pore size. As Stops (1970) pointed out, when the gas mean free path is smaller than the pore throat diameter, the motion of gas molecules is determined by their collision with each other. Gas molecules collide with the wall less frequently. During this period, viscous flow exists, which is caused by the pressure gradient between single-component gas molecules. The mass flux of viscous flow can be calculated by the Darcy law, which can be expressed as Eq. (1.6) provided by Kast and Hohenthanner [16]:

$$N_v = -\frac{\rho_g ki}{\mu_g}(\nabla p) \tag{1.6}$$

where N_v is the mass flux caused by viscous flow (kg/(m^2 s)), k_i is the intrinsic permeability of the nano capillary (m^2), p is the pressure of the nano capillary (Pa), and ρ_g is the gas density (kg/m^3).

(2) Knudsen Diffusion: When the diameter of the pore is very small, the mean free path lies relatively close to it. In this case, the collision between gas molecules and the wall becomes the dominant effect. The gas mass flux can be expressed by the Knudsen diffusion equation [17] which is shown in Eq. (1.7):

$$N_k = -M_g D_k(\nabla C) \tag{1.7}$$

where N_k is the mass flux caused by Knudsen diffusion (kg/(m^2·s)), C is the gas mole concentration (mol/m^2), M_g is the gas molar mass (kg/mol), D_k is the Kundsen diffusion coefficient (m^2/s), and D_k can be expressed as Eq. (1.8) provided by Florence et al. [18]:

$$D_k = \frac{4k_i c}{2.81708\sqrt{\frac{k_i}{\phi}}}\sqrt{\frac{\pi RT}{2M_g}} \tag{1.8}$$

where ϕ is the porosity of a single nano capillary, which is equal to 1, R is the gas constant of 8.314 $\left(\mathrm{m}^3\ \mathrm{Pa}\right)/(\mathrm{K\ mol})$, T is the temperature (K), and c is the constant. Also, Javadpour [3] proposed following expression for estimating the Knudsen diffusion coefficient.

Table 1.1 Knudsen's permeability correction factor for tight porous media

Model	Correlation factor
Klinkenberg (1941)	$b_k = 4c\lambda\overline{P}/r$
Heid et al. [19]	$b_k = 11.419(k_\infty)^{-0.39}$
Jones and Owens [20]	$b_k = 12.639(k_\infty)^{-0.33}$
Sampath and Keighin [21]	$b_k = 13.851(k_\infty/\phi)^{-0.53}$
Florence et al. [18]	$b_k = \beta(k_\infty/\phi)^{-0.5}$
Civan and Vafai [13]	$b_k = 0.0094(k_\infty/\phi)^{-0.5}$

$$D_k = \frac{d_{pore}}{3}\sqrt{\frac{8RT}{\pi M}} \tag{1.9}$$

(3) Gas slippage. Gas slippage is defined as an effect which leads the flow in the pores to deviate from viscous flow to non-laminar flow. Gas slippage occurs when the pore throat diameter is within the mean free path of gas molecules. Due to this, gas molecules will slip when they are in contact with pore surface and become fast. In low-permeability formations (less than 0.001 md) or when the pressure is very low, gas slip flow cannot be omitted such as studying gas transport in tight reservoirs. Under such kind of flow conditions, gas absolute permeability depends on gas pressure which can be expressed as follows:

$$k_a = k_\infty\left[1 + \frac{b_K}{\overline{p}}\right] \tag{1.10}$$

where k_a is the apparent permeability, k_∞ is the intrinsic permeability, $\overline{p}$ is the average gas pressure, b_K is the slip coefficient. Some empirical models have been developed to account for slip flow, such as correlations developed from the Darcy matrix permeability; correlations developed based on flow mechanisms; and semi-empirical analytical models. Table 1.1 has listed the gas slippage factor proposed by different scholars.

1.2.3 Simulation Methods for Gas Flow Through Nano Pores

Different modeling approaches have been adopted to simulate the flow of gas in nanotubes. Burnett [22] introduced the Burnett equation type method in 1935. Bird [23] and Bhattacharya and Lie [24] both tried the molecular dynamics (MD) method. Tokumasu and Matsumoto [25] and Karniadakis et al. [26] used direct simulation Monte Carlo (DSMC) to study gas flow characteristics. Hornyak et al. [27] used the Lattice-Boltzmann (LB) method to study gas flow. However, all of these modeling methods consume excessive space and time when systems are larger than a few microns, rendering them impracticable. The situation worsens when attempting to make accurate simulations with very small time steps and grid sizes, as convergence becomes a significant problem. Some researchers have attempted to derive an equation to characterize the law of gas flow. Klinkenberg (1941) introduced the Klinkenberg coefficient to consider the slip effect when gas flows in nano pores. And Beskok and Karniadakis [28] derived a unified Hagen–Poiseuille-type equation for volumetric gas flow through a single pipe. However, the applicability of these methods requires further investigation, and these modeling results have not yet been compared to real experimental data. The concept of apparent permeability for shale gas was first proposed by Javadpour [7] to simplify simulation work. In 2009, he proposed the concept of apparent permeability, considering Knudsen diffusion and advection flow. Using this method, the flux vector term can be expressed simply in the form of a Darcy equation, which greatly reduces the computational complexity. Then, Civan and Vafai [13] and Ziarani and Aguilera [12] derived the expression for apparent permeability in the form of a Knudsen number based on a unified Hagen-Poiseuille equation [28]. Shabro et al. [29, 30] applied the concept of apparent permeability further in pore scale modeling for shale gas.

1.3 Shale Permeability Measurement Methods

1.3.1 Permeability and Intrinsic Permeability

Permeability is defined as the capacity of fluid flow in the reservoir rock. It is one of the most important properties of the rock and not the fluid. Darcy law has been widely used to measure permeability which is shown as follows:

$$q = -\frac{kA}{\mu}\frac{dp}{dx} \tag{1.11}$$

where q is the fluid flow rate (cc/s), is the fluid viscosity (cP), k is the permeability (d), A is the surface area (cm^2), dp/dx is the pressure gradient along the fluid flow direction. In unconventional shale gas reservoirs, we often use the concept of intrinsic permeability. The intrinsic permeability measurement is important as it can be used to calculate many useful parameters such as gas mean free path, Knudsen number, average pore radius, and tortuosity. The intrinsic permeability is only related with reservoir properties, which does not depend on the test fluids and flow conditions.

1.3.2 Permeability Measurement Methods

Permeability is an important parameter controlling gas flow in the reservoir. Conventional steady state methods to obtain the gas permeability of shale will not be applicable due to ultra-low permeability. A long time is needed for the flow rate to stabilize. So, for unconventional shale permeability measurement, transient permeability measurement is commonly used, such as transient pressure pulse decay experiments. Many scholars have studied the methods to measure gas permeability for low permeability samples. Following are three widely used methods to obtain the shale gas permeability, such as pulse-decay Technique, crushed samples dynamic pycnometry, and canister desorption Test.

Pulse-decay technique

Pulse-decay method can be used to determine tight/shale oil/gas permeability. The intrinsic permeability can be measured from the apparent permeability and reciprocal pore pressure relationship. Alnoaimi and Kovscek have used pulse-decay method to determine the permeability of an Eagle Ford shale sample Alnoaimi and Kovscek [31]. In pulse-decay method, the apparatus consists of an upstream reservoir, a downstream reservoir, and a cell containing the rock sample with applying confining pressure. Pressure transducer is used to measure the pressure difference ($\Delta \boldsymbol{p}$) change between the upstream and downstream reservoir with time (t). Then the $\Delta \boldsymbol{p} \sim \boldsymbol{t}$ data can be analyzed to obtain the sample permeability. The permeability obtained from pulse-decay method can represent the in-situ permeability of the reservoir and is widely used in laboratories. However, this technic is dependent on the test gas. So, using the commonly used He or N_2 to measure the permeability of shale sample which is mainly composed of methane will be inappropriate.

Canister desorption test

Desorption test is generally used to evaluate the gas content of shale gas reservoirs. Also, desorption test can be used to obtain the permeability of shale. When the drill sample is obtained from the shale well, it is instantly put into the canister with reservoir temperature to desorb. During this process, the cumulative desorption gas will be recorded with time. According to the desorption data, the shale permeability can be obtained using analytical methods. Canister desorption test is generally similar with the permeability measurement with crushed sample without exerting confining pressure. This method is comparably better than the crushed sample method as the test gas is the natural gas in the reservoir.

Crushed samples dynamic pycnometry

Crushed sample can be used to measure shale properties, such as density, porosity, and adsorption isotherms. This method is generally used to measure shale permeability as it can be used on small rock particles and faster than other permeability measurement methods. Measuring shale permeability using crushed samples was designed by Luffel et al. [32]. Crushed sample can also be used to measure gas permeability of the shale sample. The methodologies of using crushed sample to measure shale gas permeability have been detailed by Cui et al. [33]. There are two reservoirs: the reference cell and the sample cell. Two pressure transduces are used to monitor the pressure in the two reservoirs. During the gas diffusing into the sample cell, the pressure data have been recorded until the equilibrium between the two reservoirs achieved. Then, the pressure vs. time data can be used to calculate the porosity and permeability of the shale sample. The permeability measured using crushed sample can only represent the properties of primary micro-pores. For more accurate results, several experiments under different pressure conditions should be conducted using this method.

1.4 Remark

In this chapter, the general concepts regarding to shale gas have been presented. Also, the importance of shale gas production, the general flow mechanisms, conventional simulation models and methods have been summarized. The experimental methods for shale gas core analysis have been provided.

This chapter is only a general introduction about the shale gas reservoir characteristics, mechanisms and models for fluid flow in shale gas reservoirs. In the following chapters, more specific models will be presented from the single component gas, two component gases, water–gas two phase transport, and production with multi-stage hydraulic fracturing horizontal well.

References

1. Sondergeld, C. H., et al. (2010). Petrophysical considerations in evaluating and producing shale gas resources. In *SPE Unconventional Resources Conference/Gas Technology Symposium*. SPE.
2. Yi, S., et al. (2023). Environmental concerns in the United States: Can renewable energy, fossil fuel energy, and natural resources depletion help? *Gondwana Research, 117*, 41–55.
3. Javadpour, F., Fisher, D., & M. Unsworth. (2007). Nanoscale gas flow in shale gas sediments. *Journal of Canadian Petroleum Technology*, *46*(10).
4. Loucks, R. G., et al. (2009). Morphology, genesis, and distribution of nanometer-scale pores in siliceous mudstones of the Mississippian Barnett Shale. *Journal of Sedimentary Research, 79*(12), 848–861.
5. Wang, F. P., et al. (2009). Pore networks and fluid flow in gas shales. In *SPE Annual Technical Conference and Exhibition?* SPE.
6. Curtis, M. E., et al. (2011). Transmission and scanning electron microscopy investigation of pore connectivity of gas shales on the nanoscale. In *SPE Unconventional Resources Conference/ Gas Technology Symposium*. SPE.
7. Javadpour, F. (2009). Nanopores and apparent permeability of gas flow in mudrocks (shales and siltstone). *Journal of Canadian Petroleum Technology, 48*(08), 16–21.
8. Jeje, A. A., & Mohtadi, F. (1984). *An introduction to the properties of fluids and solids*. University of Calgary Press.
9. Roy, S., et al. (2003). Modeling gas flow through microchannels and nanopores. *Journal of Applied Physics, 93*(8), 4870–4879.

10. Loeb, L. B. (2004). *The kinetic theory of gases.* Courier Corporation.
11. Struchtrup, H., & Struchtrup, H. (2005). *Macroscopic transport equations for rarefied gas flows.* Springer.
12. Ziarani, A. S., & Aguilera, R. (2012). Knudsen's permeability correction for tight porous media. *Transport in porous media, 91*, 239–260.
13. Civan, F., & Vafai, K. (2010). A review of approaches for describing gas transfer through extremely tight porous media. In *AIP Conference Proceedings.*
14. Yucel Akkutlu, I., & Fathi, E. (2012). *Multiscale gas transport in shales with local kerogen heterogeneities. SPE Journal, 17*(04), 1002–1011.
15. Swami, V., & Settari, A. (2012). A pore scale gas flow model for shale gas reservoir. In *SPE Unconventional Resources Conference/Gas Technology Symposium.* SPE.
16. Kast, W., & Hohenthanner, C.-R. (2000). Mass transfer within the gas-phase of porous media. *International Journal of Heat and Mass Transfer, 43*(5), 807–823.
17. Freeman, C., Moridis, G., & Blasingame, T. (2011). A numerical study of microscale flow behavior in tight gas and shale gas reservoir systems. *Transport in Porous Media, 90*, 253–268.
18. Florence, F. A., et al. (2007). Improved permeability prediction relations for low-permeability sands. In *SPE Rocky Mountain Petroleum Technology Conference/Low-Permeability Reservoirs Symposium.* SPE.
19. Heid, J., et al. (1950). Study of the permeability of rocks to homogeneous fluids. In *Drilling and production practice.* OnePetro.
20. Jones, F. O., & Owens, W. (1980). A laboratory study of low-permeability gas sands. *Journal of petroleum Technology, 32*(09), 1631–1640.
21. Sampath, K., & Keighin, C. W. (1982). Factors affecting gas slippage in tight sandstones of cretaceous age in the Uinta basin. *Journal of Petroleum Technology, 34*(11), 2715–2720.
22. Burnett, D. (1935). The distribution of velocities in a slightly non-uniform gas. *Proceedings of the London Mathematical Society, 2*(1), 385–430.
23. Bird, G. A. (1994). *Molecular gas dynamics and the direct simulation of gas flows.* Oxford University Press.
24. Bhattacharya, D., & Lie, G. (1991). Nonequilibrium gas flow in the transition regime: A molecular-dynamics study. *Physical Review A, 43*(2), 761.
25. Tokumasu, T., & Matsumoto, Y. (1999). Dynamic molecular collision (DMC) model for rarefied gas flow simulations by the DSMC method. *Physics of Fluids, 11*(7), 1907–1920.
26. Karniadakis, G. E., Beskok, A., & Gad-el-Hak, M. (2002). Micro flows: Fundamentals and simulation. *Applied Mechanics Reviews, 55*(4), B76–B76.
27. Hornyak, G. L., et al. (2008). *Introduction to nanoscience and nanotechnology.* CRC Press.
28. Beskok, A., & Karniadakis, G. E. (1999). Report: A model for flows in channels, pipes, and ducts at micro and nano scales. *Microscale Thermophysical Engineering, 3*(1), 43–77.
29. Shabro, V., Torres-Verdin, C., & Javadpour. F. (2011). Numerical simulation of shale-gas production: from pore-scale modeling of slip-flow, Knudsen diffusion, and Langmuir desorption to reservoir modeling of compressible fluid. In *SPE Unconventional Resources Conference/ Gas Technology Symposium.* SPE.
30. Shabro, V., Torres-Verdin, C., & Sepehrnoori, K. (2012). Forecasting gas production in organic shale with the combined numerical simulation of gas diffusion in kerogen, Langmuir desorption from kerogen surfaces, and advection in nanopores. In *SPE Annual Technical Conference and Exhibition*? SPE.
31. Alnoaimi, K., & Kovscek, A. (2013). Experimental and numerical analysis of gas transport in shale including the role of sorption. In *SPE Annual Technical Conference and Exhibition*? SPE.
32. Luffel, D., Hopkins, C., & Schettler, P., Jr. (1993). Matrix permeability measurement of gas productive shales. In *SPE Annual Technical Conference and Exhibition*? SPE.
33. Cui, X., Bustin, R. M., & Dipple, G. (2004). Selective transport of CO_2, CH_4, and N_2 in coals: Insights from modeling of experimental gas adsorption data. *Fuel, 83*(3), 293–303.

Chapter 2
Pore Structure Characteristics of Shale

Shale gas reservoirs are characterized as organic-rich depositions with extremely low matrix permeability and clusters of mineral-filled "natural" fractures. Shale gas reservoirs are composed of fine-grained sedimentary rocks with permeability in the nano scale. Natural gas can be stored in shale reservoirs as free gas in nano pores and fractures, adsorbed gas on the surface of nano pores, and dissolved gas in the kerogen. Many geologic and reservoir properties such as depositional environment of shale, depth, structure, total organic content, and thermal maturity are used to provide an overview of the geologic characteristics of the major shale gas and shale oil formations. Also, these parameters can be used to select the shale gas basins and formations which are worth exploiting. An important factor among these parameters is the depositional environment of the shale (marine or non-marine). Marine-deposited shales tend to have lower clay content and tend to be high in brittle minerals such as quartz, feldspar, and carbonates. Brittle shales respond favorably to hydraulic stimulation. Shales deposited in non-marine settings (lacustrine, fluvial) tend to be higher in clay, more ductile, and less suitable for hydraulic stimulation.

In the Silurian shale of the Sichuan Basin, Qiongzhusi and Longmaxi shale plays are two most important marine shales for natural gas exploration and development in China. Microstructures of shales exhibit a high degree of complexity and heterogeneity. Nanometer- to micrometer- scale pore systems have been found in the organic matter and matrix of inorganic grain, which have significant impact on gas generation and gas storage. To study the complex pore structure of shale, various methods have been developed to investigate the pore structure of shale, such as fluid invasion methods, optical methods, and computer aided methods. Fluid invasion methods include mercury intrusion capillary pressure (MICP), helium pycnometer, etc., which can obtain an estimation of petrophysical properties of shales, including porosity, specific surface area, and pore size distribution. Optical methods include thin section imaging, scanning electron microscopy (SEM), transmission electron microscopy (TEM), and scanning transmission x-ray microscopy, which can provide an estimation of pore size, pore structure characteristics, morphology, and pore connectivity.

C. Guo and Z. Yang, *Shale Gas Production: Concept, Models, and Techniques*, Petroleum Engineering, https://doi.org/10.1007/978-3-031-84552-9_2

By ion beam milling, the imaging surface can be polished so that the imaging quality can be greatly improved, such as focused ion beam- SEM (FIB-SEM) which can provide gallium ions to strike sample surface to obtain a cross-section, and atomic force microscopy (AFM) which uses the argon ion beam to polish the surface. This technology can avoid the damage from cutting and polishing. Using this technology, Javadpour [1] have detected the nano pores in the Barnett shale sample [1]. Passey et al. [2] used this method to show the morphology of pores. Sondergeld et al. provided the kerogen percentage change during the polishing and the surface feature of Floyd shale, Barnett shale and Fayettecille shale [3, 4]. Computer aided methods such as X-ray Computed Tomography, 3D reconstruction, and digital rock, which can provide 3D digital rock by combining the imaged 2D slices. After cross-sections are sliced sequentially using FIB and high resolution images for each slice is obtained using SEM, image processing software (such as ImageJ) can be used to construct the 3D model by combining the series of images obtained. The 3D model can be used to calculate porosity, simulate fluid flow through nano pore throats, and analyze pores and mineral distribution. Elgmati et al. [5] reconstructed a 3D model for Fayetteville shale to analyze porosity. The same method was used by Passey et al. [2] to reconstruct 3D porosity model and to investigate the porosity distribution within the organic matter. Sondergeld et al. [4] constructed a 3D porosity network for Barnett shale by 3D reconstruction.

Based on recent imaging and experiment methods, several pore types have been detected in the shale rocks, such as organic pores, interparticle mineral pores, microfractures, micro fractures, intra-crystalline, elongated pores between similarly oriented clay sheets, crescent shaped pores in saddle reefs of folded sheets of clay, and large jagged pores, etc. Schieber [6] classified pores into phyllosilicate framework pores, carbonate dissolution pores, and organic matter pores. Milner et al. [7] added intraparticle pores based on the Schieber's research. Six pore types were identified by Slatt and O'Brien [8] in the Barnett and Woodford shales, including porous floccules, organo-porosity, fecal pellets, fossil fragments, intraparticle grains/pores, micro-channels, and micro-fractures. Generally, these pore types can be classified into two categories: (a). organic matter related pores, and (b). inorganic matter related pores include phyllosilicate framework pores, carbonate dissolution pores, and intra-particle pores, pyrite framboids and clay. Formation porosity, total organic carbon, and thermal maturity have significant effect on shale gas generation, which determines whether the target area is available for economical developing. Mineral composition such as quartz content and brittle minerals determines whether the target area is suitable for applying hydraulic fracturing. Comparison between China shale plays with the mature shale plays in the North America will facilitate us to have a more accurate understanding about China shale, which contributes to the reserve assessment and production plan determination. A lot of work have already been conducted for the pore structure characterization since the introduction of high resolution imaging system. However, rare work has been conducted for the Qiongzhusi and Longmaxi shale by comparing with mature shale plays in North America from pore structure, porosity, TOC, thermal maturity, and mineral composition.

In the chapter, we have focused on the pore structure characteristics and petrophysical properties of two China marine shale plays in the Sichuan Basin, China. Firstly, we have studied the pore structure analysis using SEM imaging, MICP, and 3D reconstructed model. Then, we studied the clay mineral structure using SEM and the clay mineral composition using X-ray diffraction test to determine whether the target area is suitable for applying hydraulic fracturing. Lastly, we have compared the China shale plays with other five mature shale plays in the North America based on porosity, TOC, quartz content, thermal maturity, and etc. The findings from this study can contribute to the understanding of formation characteristics and reserve assessment of China shale plays. Also, it will facilitate the development of suitable and efficient production plan for China shale plays.

2.1 Pore Structure Characterizing Methods

2.1.1 Porosity Measurement

For the shale porosity measurement, gas expansion method is widely used. Helium is commonly used due to the tiny molecular diameter which can go through narrow pores and low absorptivity on shale surface. In this study, helium porosimeter was used for direct grain volume and pore volume measurement in an auxiliary cell at isothermal conditions. Subsequently, porosity and grain density can be derived from the direct measurements. The helium porosimeter includes sample chamber, matrix cup (grain volume cell) for one-inch diameter samples with five calibrated disks for system calibration, integrated pressure transducer, reference cell and gas flow manifold assembly, digital data read-out, thermometer and associated plumbing.

2.1.2 MICP

Mercury Injection Capillary Porosimeter (MICP) monitors the volume of Hg that is injected into a dried core sample at each pressure level. Mercury is nonwetting phase and it can only access to interconnected pores. Therefore, the porosity calculated by this method is effective porosity. Based on Hg porosimetry, the porosity of the shales mainly depends on the degree of pore volume development in pores less than 10 μm. A lot of research has used this method to analyze the pore size distribution of the rock. Pore-throat sizes range from about 2 to 0.03 mm in tight-gas sandstones, and range from 0.1 to 0.005 mm in shales. Devonian-Mississippian (D-M) and Jurassic shales were identified by Hg porosimetry. It is found that the total porosity increases with decreases of the Si/Al ratio, which is a reflection of the porosity associated with clay mineral phases. The same method was used to detect the pore size distribution and porosity for Utica shale and Haynesville shale, Norwegin North Sea shale, Cretaceous shale, Barnett shale, and Devonian shale. Ramirez and Morelli [9] utilizes this technique to determine pore size distribution and surface (pore) areas of spent oil shale. Comisky et al. [10] measured MICP profiles on tight Eagle Ford shale with a variety of sample sizes to optimize the rock preparation and data reduction workflow for determining the storage properties of the rock, particularly porosity. Pore structure versus sediment texture relationship was examined by 28 sedimentary rock samples from Northwest Territories by using MICP.

2.1.3 Mineralogy Analysis

It is important to conduct mineralogy analysis as the mineral composition affects the brittleness of shale and the hydraulic fracturing design. The clay minerals will absorb water in the process of drilling and fracturing which results in wellbore collapse and low return rate of the fracturing fluid. In addition, the clay minerals in the pores can absorb gas which hinders gas production.

X-ray diffraction is a versatile, non-destructive technique that reveals detailed information about the chemical composition and crystallographic structure of natural and manufactured materials. It provides qualitative results of mineralogy and clay mineralogy which influence pore space and formation properties which are important in the gas production process. X-ray diffraction (XRD) analysis was performed on powered shale using Multi-Purpose diffractometric (MPD). The Philips X-Pert MPD is for powder or polycrystalline diffraction analysis at room or non-ambient temperatures with controlled atmospheres. This device provides both qualitative and qualitative results of mineralogy and clay mineralogy which influence pore space and formation properties. Normal operating power is 40 kV, 50 mA (2.0 kW).

As the XRD method detect the intensity of each mineral, some minerals or clay minerals with small amount have very weak intensity which is hard to observe the peak in the XRD curve, the mineralogy and the clay mineralogy are both utilized in this study for more accurate measurement of the clay mineralogy and non-clay mineralogy.

2.1.4 FIB-SEM Imaging

FIB-SEM is a high-resolution imaging technology. Two SEMs were used in this research: FEI Helios NanoLab 600 and 650, both of which have dual beam system (SEM/Focused ion beam). Helios has three detectors to detect different signals from electron specimen interactions and two accessories for advanced research of sample. They are backscattered electrons detector, secondary electrons detector, X-ray detector, Maps software and Auto slides and view software. Thus, SEM has two modes. Backscattered mode is detecting the backscattered electrons to provide information about the distribution of different elements because that mode is strongly related to the atomic number of the elements. The higher the gray level, the lower the atomic number of the element is. The kerogen can be found easily under backscattered mode. Secondary mode is the normal mode which is used to provide high resolution images that closely approximate the real world by detecting secondary electrons. Using SEM, the compositional and microstructural features of the sample surface can be observed at micro and nano-scale.

Unlike regular electron microscope, FIB-SEM is inherently destructive to the specimen as high beam currents of FIB is used for site specific sputtering or milling. Combined use of SEM and FIB can reconstruct the 3D model which can show the porosity, the pore connection, and also mineral distribution of the target area of the sample.

2.1.5 *Confocal Microscopy Imaging*

The conventional pore identification methods only demonstrates the pore characteristics in 2D, and it is impossible to use it to analyze pore structures in 3D, or the pores ranging from several microns to nanometer. Therefore, the confocal microscope method is used to solve the problem. It was first brought to petroleum industry by Li and Wan [11] to investigate the equilibrium asphaltene particle shape, particle size, and size distribution, and then was used to describe the characteristics of brine-in-bitumen emulsions. Confocal microscopy was used to identity micro fractures and gas desorption [12], and to efficiently and accurately measure pore-body and pore-throat size distributions, and to predict porosity and permeability of carbonate rocks [13].

In a confocal laser scanning microscope, a laser beam passes through a light source aperture and then is focused by an objective lens into a small focal volume within or on the surface of a specimen. As the laser is scattered and reflected by the beam splitter and scanned across the sample in a defined focal plane, secondary fluorescence emitted from points on the specimen pass back through the beam splitter and then focus as a confocal point at the detector pinhole aperture. After passing the pinhole, the light intensity is detected by a photo detection device (usually a photomultiplier tube (PMT) or avalanche photodiode), transforming the light signal into an electrical one that is recorded by a computer.

All above experimental methods can be used to investigate porosity and pore structure. For different pore scale, different experimental methods should be used. Figure 2.1 shows the methodologies commonly used for investigating porous media. It can be found that Transmission electron microscope and Helium porosimeter can detect nano-scale pores. Some imaging methods can only detect pores larger than 100 nm, such as the micro-CT, confocal microscope, and thin section detection. The advantages and disadvantages of the methods to measure porosity and pore structure have been listed in the Table 2.1.

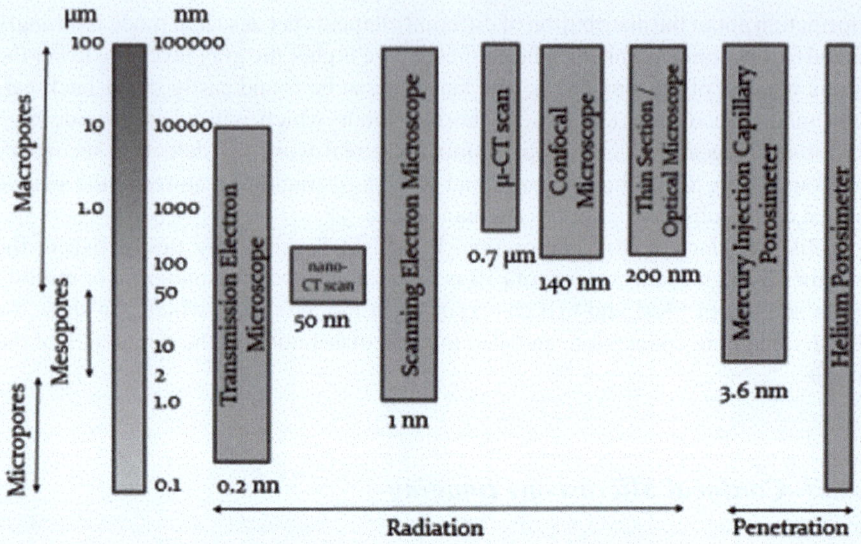

Fig. 2.1 The methodologies commonly used for investigating porous media

Table 2.1 The advantages and disadvantages of methods to measure porosity and pore structure

Methods	Advantage	Disadvantage
Gas expansion	Accurate	Long diffusion time
Thin section	Heterogeneous, detect pore connection	Low resolution
MICP	Connected pores, high resolution	Destroys sample, nonrepeatable, sample must be disposed of safely
µ-CT	Measures pore throat distribution, heterogeneous, 3-D	Low resolution, 9 µm
SEM	High-resolution, surface feature	Small area
SEM maps	Large area	Low resolution
TEM	High resolution, < 1 nm, detect pore connection	Small thickness, < 100 nm
Nano-CT	High resolution, 50 nm, 3-D	Small sample
FIB/SEM	High resolution, 1 nm, 3-D	Long time cost

2.1.6 Organic Geochemistry

The main difference between shale and other reservoir rocks is the presence of total organic carbon (TOC). The organic matters not only generate gas, but also provide space for gas storage. Most of organic matters are located in the formation independently and disconnected with each other. The pore network within the organic matters and other type of pores existing in the rock are important for gas storage capacity analysis and treatment design. Thermal maturity is an indicator of pressure and temperature which the rock had been subjected to. It also indicates whether oil or gas has been generated during the process. Vitrinite reflectance (Ro) is a commonly used as thermal maturity indicator. Thermal maturity comprises oil (0.6% < Ro < 0.8%), wet-gas (0.8% < Ro < 1.1%), and dry-gas windows (Ro > 1.5%).

2.2 Pore Structure Characteristics of Shale

2.2.1 Pore Types Analysis Based on Imaging Methods

SEM methods can be used to effectively analyze the relatively large micron-scale pores. This research has been conducted to analyze the different pore types developed in Longmaxi shale gas reservoir, which includes primary intergranular pores, intraparticle pores, organic pores, residual intergranular pores, dissolution pores, and intergranular pores. Combined with previous analysis brought up by other researchers, the results showed that the majority distributed in residual intergranular pores, secondary dissolution pores, and intercrystalline pores, but less in primary intergranular pores. It is considered that the primary intergranular pores disappeared before the natural gas enrichment, intense compaction and cementation which lead to formation tight. Here, one shale sample L415 from Longmaxi shale is presented. The SEM results for shale sample L415 are shown in Fig. 2.2.

The SEM analysis shows that the pore diameter of reservoir are mainly between 1 and 100 μm. The radiuses of primary intergranular pores and residual intergranular pores are larger, generally between 50 and 100 μm, and the residual intergranular pores are formed of either the overgrowth of clastic particles or the clay minerals filling in original intergranular pore. Dissolution pore radius is generally less than 50 μm, and is composed of a series of honeycomb like pores with different sizes. The intergranular pore is relatively small, generally less than 10 μm, and mainly exists in kaolinite, illite, chlorite and some other authigenic clay minerals. The microstructure of the organic matters and pyrite are shown in Fig. 2.2, from which the pores presented are in nano scale.

2.2.2 Pore Characteristics Analysis with 3-D Modeling

To analyze the nanopores more accurately, a 3-D model pore analysis method was used during this study. The main principle of this method is milling the sample (L416) surface (Fig. 2.3) using a focused ion beam (FIB). The FIB is generated by the ion source, accelerated, and focused into a high-current ion beam. Then, the internal section of the region of interest can be observed under scanning electron microscopy. Further, FIB can mill a series of cross sections with small intervals.

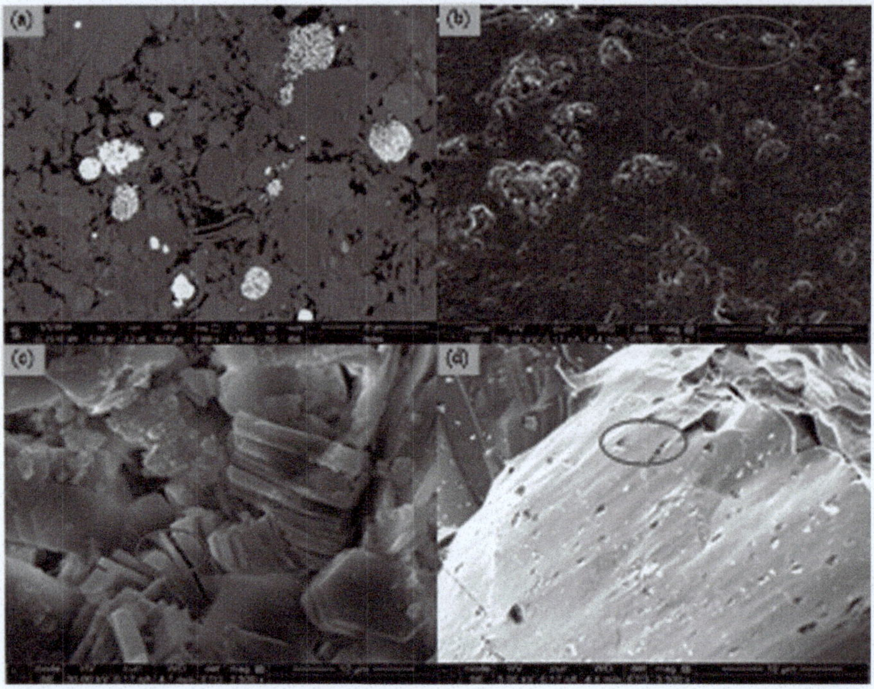

Fig. 2.2 Pore types of Longmaxi shale sample L415: **a** primary intergranular pore, 100 μm; **b** dissolution pore, < 40 μm; **c** kaolinite intercrystalline pores, < 10 μm; **d** dissolution pore, < 10 μm

Thus, reconstructing a 3-D model of pore connectivity can be analyzed by FIB/SEM to obtain a series of continuous cross section images. It has been used to analyze the pore connections inside the shale sample. However, since FIB is only suitable for analyzing a small area, the connected pore throat of a 3-D model of shale sample cannot be reconstructed because the residual intergranular pores and dissolution pores are usually larger than 50 microns. After using image-processing software, the 3-D model was reconstructed from this series of continuous cross section images. In contrast to the mathematical model, the synthetic model is completely consistent with the internal structure of the sample. Since the cross section is strongly milled by an electron beam, the pore structure of the cross section can represent the real pore structure of the sample. A 3-D model of the pore network of the organic pores for sample L416 is as shown in Fig. 2.4a and the pore size distribution is shown in Fig. 2.4b.

Figure 2.5 is the reconstructed 3-D model of quartz and kaolinite from a series of continuous particle cross sections for sample L416. The process of building the internal structure model of the quartz particles (Fig. 2.5a) is the same as the process used for the 3-D model construction of kaolinite (Fig. 2.5b). From the experimental

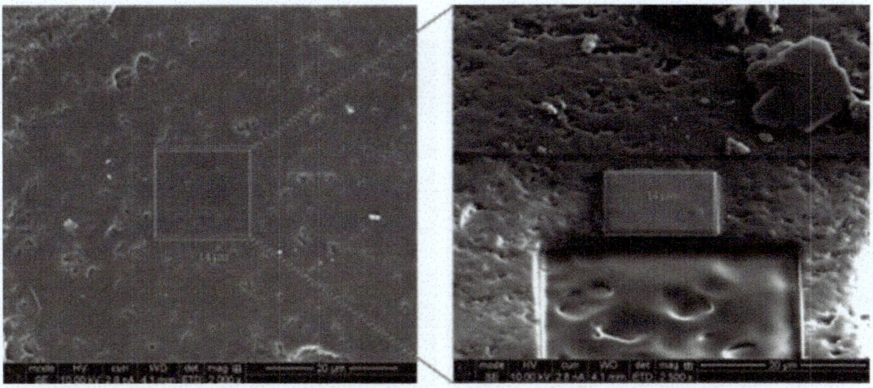

Fig. 2.3 Mill the surface of the area of interest with FIB for sample L416

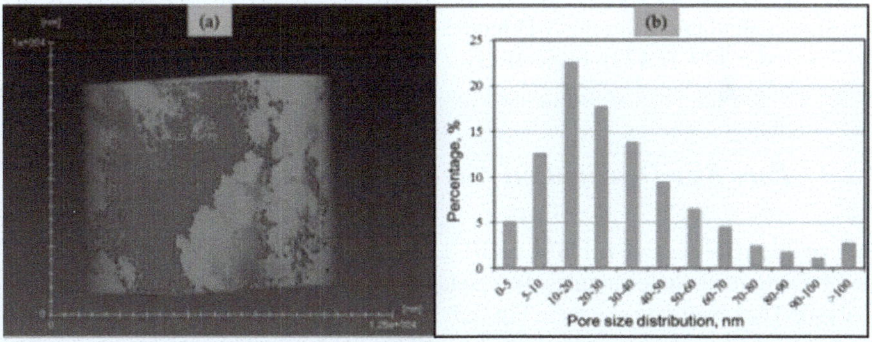

Fig. 2.4. 3-D model reconstructed and pore size distribution for sample L416

analysis, only a few of the quartz particles have isolated nano pores in the Longmaxi shale sample L416, and the pore connectivity is too poor to form the effective porosity. Because of the inconsistency of the orientation of clay, nanoscale porosity rarely occurs in kaolinite, and the effective porosity is not developed. The nanoscale pores exist in the clastic particles, and the clay minerals are mostly isolated and disconnected. According to the results of a variety of experimental analyses, the Longmaxi shale sample is not only filled by micron pores but also nanoscale pores. Overall, the pore radii are less than 100 μm, in a range from 100 microns to nanometer scale. The pores between the clastic particles and the clay minerals are micron pores, which can connect with each other to form effective pore network. However, the nanopores in the minerals, such as the grains of quartz and kaolinite particles, are isolated, which leads to difficulty in building effective pore network. Analyzing the CT results with the e-core software, the pore network can be reconstructed for four selected shale samples (L416, L421, L424, and L427) as shown in Fig. 2.6. The

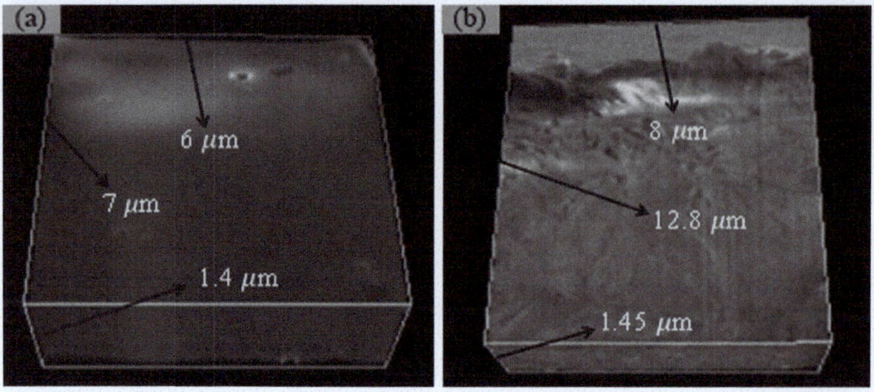

Fig. 2.5. 3-D model of the pore connection of quartz particles **a** and kaolinite **b** for sample L416

porosities calculated from the 3-D model are 6.34% (L416), 2.43% (L421), 5.04% (L424), and 6.32% (L427), respectively.

2.2.3 Pore Size Distribution Analysis Based on MICP

Ten samples (L401, L404, L406, L410, L414, L418, L420, L422, L425, L428) of the 36 Longmaxi shale samples were selected to conduct the pore size distribution analysis. The results are shown in Fig. 2.7. According to the results, most of the pores are nano pores, mainly ranging from 5 to 20 nm. The detected pores are connected pores. They may play a role as channels for shale gas flow though.

2.3 Clay Mineral Analysis Based on SEM and XRD Test

2.3.1 Clay Mineral Structure Based on SEM

The SEM provides the image of the structure of clay minerals (Fig. 2.8). The clay minerals presented in the Longmaxi shale sample L417 are mainly kaolinite, illite, and chlorite.

Kaolinite not only reduces the porosity of the rock but also separates the original intergranular pores into many micropores, reducing the permeability of the rock. Chlorite and illite usually adsorb to the surface of the pores or particles. Although chlorite controls the quartz enlargement to keep the original intergranular pores, it will occupy the space of the pore throat and reduce the permeability of the rock. The presence of illite also blocks the pores and reduces the permeability of the rock.

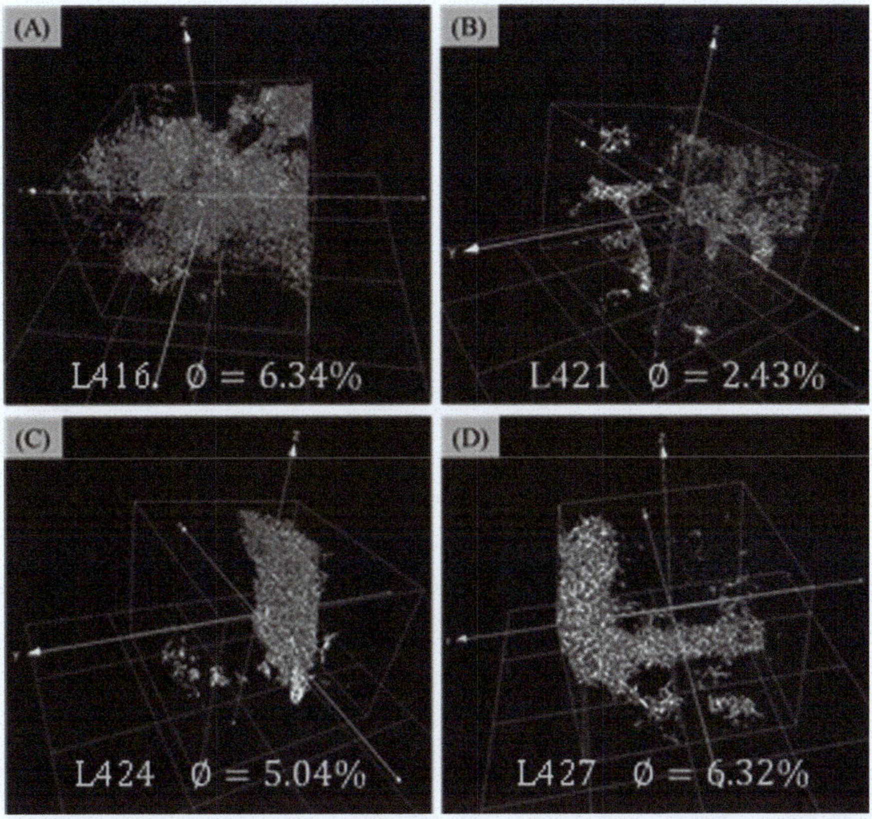

Fig. 2.6 The pore network of samples (pores in red and the pore throat in white): **a** the pore network of Longmaxi shale sample L416, **b** the pore network of sample L421, **c** the pore network of sample L424, and **d** the pore network of sample L427

2.3.2 Clay Minerals Analysis Based on XRD Test

X ray diffraction is often used in mineralogy and clay mineralogy quantitative analysis, because of its higher accuracy compared with thin section method. Due to the high content of quartz, it is difficult to detect the clay minerals at a lower content level. Therefore, to obtain a more accurate result, the clay mineral analysis experiment is taken as an additional experiment to detect the clay mineral composition and then used in the calculation of the content of each mineral. Taking the core sample form Longmaxi shale as an example, the XRD result of sample L417 is shown in Fig. 2.9. The red curve is the raw mineral analysis result. From this curve, it can be found that sample L417 contains a lot of quartz and a small part of kaolinite and calcite. To separate the mixture, dissolving the powder sample in water, the clay particles suspend in the upper solution and the mineral particles start to precipitate at the bottom according to the different settling velocity between mineral particles and

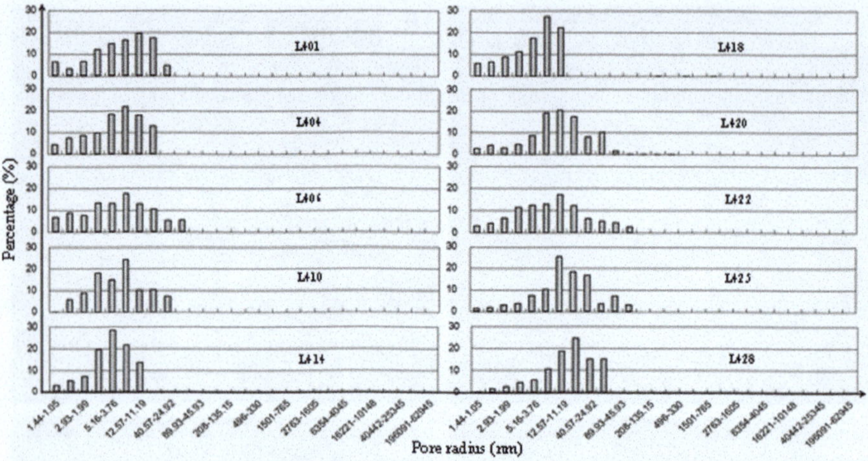

Fig. 2.7 Pore size distribution of ten selected Longmaxi shale sample using MICP method

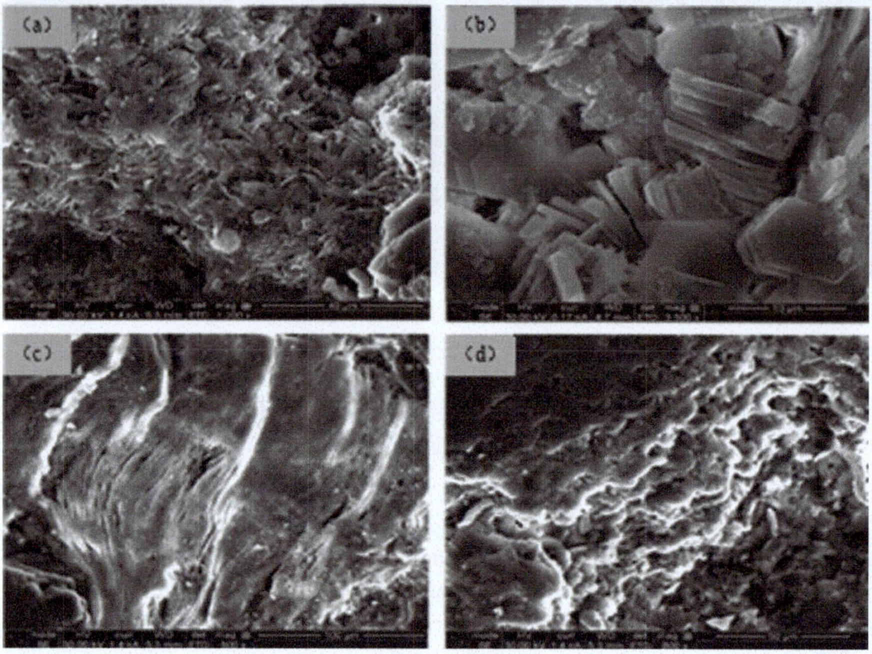

Fig. 2.8 Clay mineral structure in the Longmaxi shale sample L417: **a** vermiform kaolinite, **b** kaolinite booklet, **c** lamellar chlorite, and **d** fibrous illite

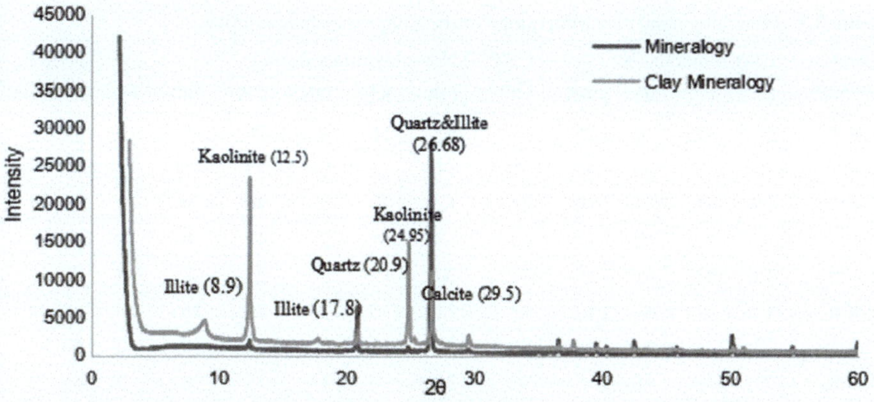

Fig. 2.9 The XRD result of sample L417 in Longmaxi shale

clay particles. After a period of time, the upper part of the solution will be taken out, dried, and then used to do the clay mineral analysis. Since most of the quartz particles has been removed, the measurement of clay mineral composition will become more accurate. In Fig. 2.9, the green curve shows the new clay minerals contents after the removing of most of the quartz particles. Comparing two curves, illite is detected in clay mineralogy experiment. XRD analysis shows that the clay minerals of Longmaxi shale reservoir mainly consist of illite and kaolinite, then followed by chlorite. The content of kaolinite is from 7.2 to 16.87%, and the average is 12.21%. The content of illite is from 8.03 to 26.8%, and the average is 16.1%.

More specific information of the clay mineralogy and non-clay mineralogy of the Longmaxi samples are shown in Table 2.2. According to the results, the total clay content is from 3 to 76%, with an average of 34%. The clay minerals usually contain chlorite, illite, and illite–smectite mixture. The illite and illite–smectite mixture have higher content which are 37% and 41% respectively, while the content of chlorite is 22%. When drilling fluid or fracturing fluid being injected into the formation, the illite–smectite mixture get swelled, and then reduce the porosity and permeability of reservoir by blocking pore throat. However, clay minerals have positive effect on shale gas adsorption and storage. Non-clay minerals mainly include quartz, K-feldspar, calcite, dolomite, anorthose, and pyrite. The quartz has the highest content value followed by calcite, which are 29% and 22% average respectively. The content of pyrite and anorthose are less than 10%. Hydraulic fracturing technology is the key to obtain successful production of shale gas reservoir. The brittleness of rock is very important for hydraulic fracturing, and the content of brittle minerals in the shale can influence the formation of induced fractures. With increasing of quartz or calcite, the brittleness of the rock will correspondingly increase. The average content of brittle minerals of these Longmaxi formation samples is 66%, which is more than 50%, so this formation is suitable for conducting hydraulic stimulation.

For the Qiongzhusi shale, seven samples were collected from a depth of − 2816 to − 3036 m. The clay mineralogy and non-clay minerology of the seven samples

Table 2.2 Mineralogy and clay mineralogy of the Longmaxi shale samples

Sample number	Clay minerals (%)				CRT (%)	Non-clay minerals (%)						NCRT (%)
	C	I	I/S	%S		QA	KF	AN	CA	DO	P	
L401	81	11	8	25	3	65	11	0	16	5	0	97
L402	20	50	30	30	35	14	5	6	33	4	3	65
L403	26	41	33	30	32	12	0	0	46	9	1	68
L404	18	42	40	10	23	17	1	2	49	8	0	77
L405	18	41	41	15	30	20	12	2	25	8	3	70
L406	18	41	41	10	32	22	0	2	28	16	0	68
L407	22	38	40	8	34	25	2	0	27	12	0	66
L408	21	38	41	5	34	24	0	1	33	8	0	66
L409	20	42	38	8	35	20	8	2	28	6	1	65
L410	25	45	30	6	36	22	6	1	28	6	1	64
L411	22	46	32	5	38	21	3	2	32	3	1	62
L412	21	40	39	10	41	25	0	2	27	5	0	59
L413	25	40	35	5	33	35	0	2	29	1	0	67
L414	28	39	33	5	28	38	0	5	25	4	0	72
L415	23	42	35	12	30	30	6	3	26	3	2	70
L416	21	40	39	5	35	33	1	3	20	7	1	65
L417	20	38	42	5	36	35	2	3	19	5	0	64
L418	20	40	40	8	33	37	0	3	22	5	0	67
L419	24	35	41	5	35	31	0	4	22	7	1	65
L420	30	40	30	8	30	33	8	7	16	3	3	70
L421	28	42	30	5	36	30	6	8	13	6	1	64
L422	25	44	31	10	41	35	0	9	10	5		59
L423	20	30	50	12	34	35	8	6	14	1	2	66
L424	15	35	50	16	37	40	3	9	8	2	1	63
L425	12	28	60	8	63	22	0	6	5	4	0	37
L426	8	26	66	8	75	18	0	3	3	1	0	25
L427	1	28	71	10	76	10	0	3	5	8	0	24
L428	15	30	55	12	46	36	0	2	10	6	0	54
L429	16	32	52	5	42	42	2	5	6	2	1	58
L430	15	38	47	22	13	37	0	4	28	16	2	87
L431	22	31	47	18	22	30	7	3	22	13	3	78
L432	24	30	46	21	18	35	7	6	24	8	2	82
L433	20	32	48	32	30	28	1	5	25	10	1	70
L434	23	32	45	45	26	37	1	1	20	12	3	74

(continued)

Table 2.2 (continued)

Sample number	Clay minerals (%)				CRT (%)	Non-clay minerals (%)						NCRT (%)
	C	I	I/S	%S		QA	KF	AN	CA	DO	P	
L435	32	30	38	58	21	26	5	3	27	16	2	79
L436	42	32	26	60	16	24	0	1	31	18	10	84

Note C Chlorite, *I* Illite, *S* Smectite, *I/S* Illite- Smectite, *%S* Percentage of Smectite layers in interstratified I/S; *CR* the ratio of total clay minerals in the whole minerals, *QA* Quartz, *KF* K-feldspar, *AN* Anorthose, *CA* Calcite, *DO* Dolomite, *P* Pyrite, *NCR* the ratio of total non-clay minerals in the whole minerals

Table 2.3 Mineralogy and clay mineralogy of the Qiongzhusi shale samples

Sample number	Clay minerals (%)				CRT (%)	Non-clay minerals (%)						NCRT (%)
	C	I	I/S	%S		QA	KF	AN	CA	DO	P	
Q401	2	96	2	20	4	90	3	1	1	1	0	96
Q402	16	80	4	22	10	80	4	2	2	1	1	90
Q403	4	93	3	25	19	71	7	1	1	0	1	81
Q404	4	95	1	26	20	70	6	2	1	1	0	80
Q405	3	93	4	18	15	80	2	2	1	0	0	85
Q406	1	93	6	21	25	70	2	2	1	0	0	75
Q407	15	80	5	23	18	73	5	2	2	0	0	82

are listed in the Table 2.3. According to the results, the total clay content is from 4 to 25%, with an average of 16%. The clay minerals are mainly illite. For the non-clay minerals, quartz has the highest content, followed by K-feldspar, with averages of 76% and 4%, respectively. The results indicate that Qiongzhusi shale is suitable to conduct hydraulic fracturing. Compared with the Longmaxi shale, the Qiongzhusi shale is more brittle and easier to be fractured.

2.4 Total Organic Carbon

Total organic carbon (TOC) content is one of the key parameters to evaluate shale gas. High TOC is beneficial for shale gas generation and enrichment. Also, it is the basis for the organic pores formation. With TOC increasing, the adsorption capability of shale increases. In this study, 36 samples were collected from – 1687 to – 2063 m in the Longmaxi shale reservoir and the relationship between TOC and depth is shown in Fig. 2.10. The average TOC is 1.29%, the maximum value is 5.05% and the minimum value is 0.2%. It has been observed that 19.4% samples contain organic carbon less than 0.5%, while 41.7% of them are more than 1%, which are buried from – 1945 to – 2045 m. Most of the samples have their TOC more than 0.5%, which means Longmaxi formation has a potential to produce shale gas.

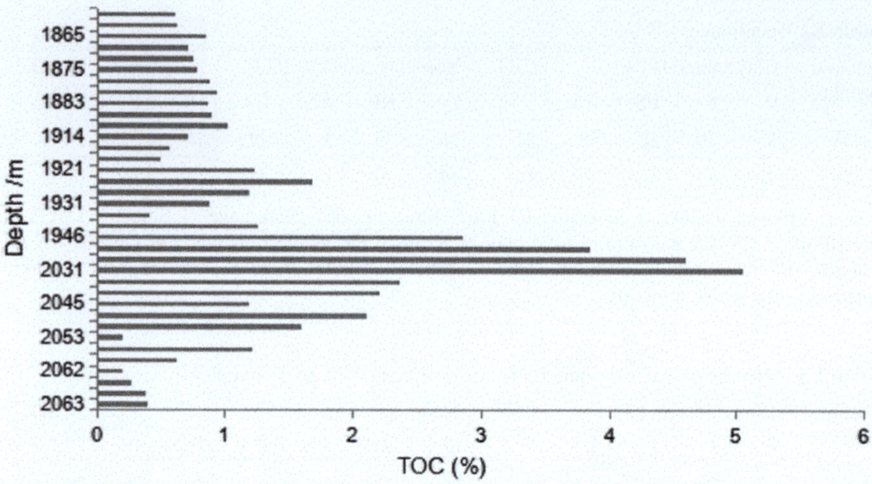

Fig. 2.10 The relationship between TOC and depth for the Longmaxi shale samples

For the Qiongzhusi shale, the relationship between TOC and depth of the seven samples is presented in the Fig. 2.11. From Fig. 2.11, it can be found that the average TOC of seven measured samples is 1.27%. The maximum value is 2.4% and the minimum value is 0.8%. A majority of samples have more than 0.5% TOC, meaning that the Qiongzhusi formation has shale gas production potential. Compared to the Longmaxi shale, the average TOC in the Qiongzhusi shale is higher. However, the TOC of the Longmaxi shale has a larger range than that of the Qiongzhusi shale, which suggests that the Longmaxi shale has large TOC layers which can produce excellent shale gas.

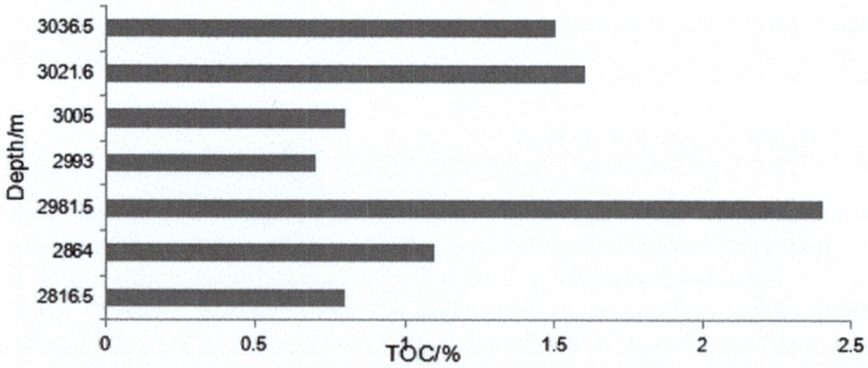

Fig. 2.11 The relationship between TOC and depth of the Qiongzhusi shale samples

2.5 Thermal Maturity

Thermal maturity is another key geochemistry parameter controlling organic hydrocarbon generation, organic matter pore development, and methane stabilization. Due to the time needed to test the thermal maturity, seven samples from 36 Longmaxi shale samples and seven samples Qiongzhusi shale samples were selected to measure their thermal maturity (Vitrinite reflectance Ro) in this study, and the results are shown in Figs. 2.12 and 2.13.

Figure 2.12 shows the relationship between thermal maturity and depth of the seven selected Longmaxi shale samples. From Fig. 2.12, it can be found that the vitrinite reflectance (Ro) of each sample is more than 2.0%, which indicates that it is

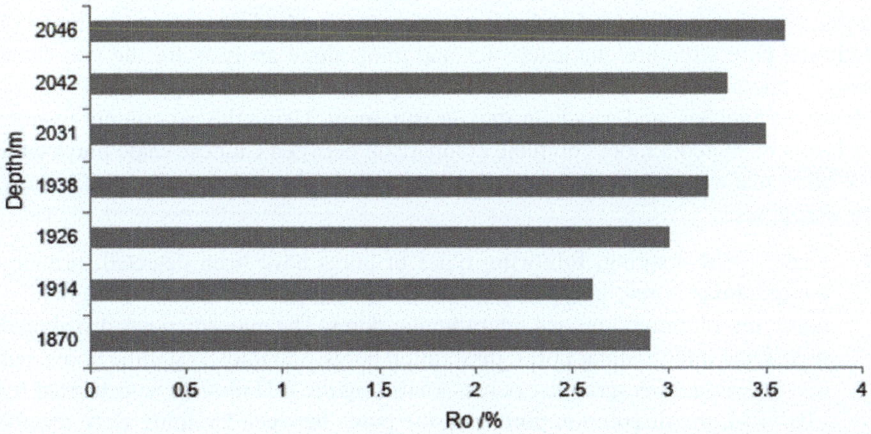

Fig. 2.12 The relationship between thermal maturity and depth of the Longmaxi shale samples

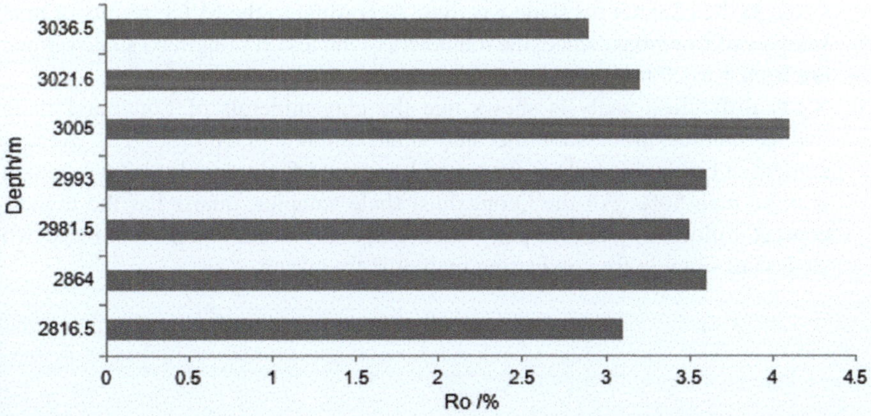

Fig. 2.13 The relationship between thermal maturity and depth of the Qiongzhusi shale samples

high enough to generate shale gas in the Longmaxi shale formation. The average Ro of the seven Longmaxi shale samples is 3.15%. The maximum value of Ro is 3.6% and the minimum valve is 2.6%.

Figure 2.13 shows the relationship between thermal maturity and depth of the seven Qiongzhusi shale samples. From Fig. 2.13, it can be found that the vitrinite reflectance (Ro) of Qiongzhusi shale samples is higher than those of Longmaxi shale samples. The average Ro of the seven Qiongzhusi shale samples is 3.43%. The maximum value of Ro is 4.1% and the minimum valve is 2.9%.

2.6 Remark

In this chapter, we presented multiple experimental methods have been deployed to study the pore structure characteristics and mineralogy analysis for the two shale plays in the Sichuan basin, such as scanning electron microscopy, MICP, X-ray diffraction analysis, and geochemistry measurement. Using the experiment results, we have conducted a more complete comparison between Chinese shale plays with five other mature shale plays in the North America. The main results of this study are as follows:

(1) Under SEM imaging, following types of pores have been detected: primary intergranular pores, intraparticle pores, organic pores, residual intergranular pores, dissolution pores, and intergranular pores. The majority pores distributed in residual intergranular pores, dissolution pores, and intercrystalline pores, but less in residual intergranular pores. Analyzing the 3-D model reconstructed by FIB/SEM, the dissolution pores and the pores between kaolinite were mostly isolated and disconnected.
(2) The nano-pores presented in the organic matters is the most important space for shale gas storage. Some other pores, such as pores presented in the pyrite, play a role as the channel for shale gas flow. According to the MICP results of nine samples of Longmaxi shale, the diameters of the tested Longmaxi shale sample are from 1 to 20 nm.
(3) X-ray diffraction analysis shows that the clay minerals of Longmaxi shale mainly include illite, kaolinite, and small amount of chlorite. The average content of brittle minerals of the tested Longmaxi shale samples is 66%, which is more than 50%. For the Qiongzhusi shale smaples, quartz has the highest content, followed by K-feldspar, with averages of 76 and 4%. Both of these two shales are suitable for conducting hydraulic fracturing.

References

1. Javadpour, F. (2009). Nanopores and apparent permeability of gas flow in mudrocks (shales and siltstone). *Journal of Canadian Petroleum Technology, 48*(08), 16–21.
2. Passey, Q. R., et al. (2010). From oil-prone source rock to gas-producing shale reservoir–geologic and petrophysical characterization of unconventional shale-gas reservoirs. In *SPE International Oil and Gas Conference and Exhibition in China*. SPE.
3. Sondergeld, C. H., et al. (2010). Petrophysical considerations in evaluating and producing shale gas resources. In *SPE Unconventional Resources Conference/Gas Technology Symposium*. SPE.
4. Sondergeld, C. H., et al. (2010). Micro-structural studies of gas shales. In *SPE Unconventional Resources Conference/Gas Technology Symposium*. SPE.
5. Elgmati, M., et al. (2011). Submicron-pore characterization of shale gas plays. In *SPE Unconventional Resources Conference/Gas Technology Symposium*. SPE.
6. Schieber, J. (2010). Common themes in the formation and preservation of intrinsic porosity in shales and mudstones–illustrated with examples across the Phanerozoic. In *SPE Unconventional Resources Conference/Gas Technology Symposium*. SPE.
7. Milner, M., McLin, R., & Petriello, J. (2010). Imaging texture and porosity in mudstones and shales: Comparison of secondary and ion-milled backscatter SEM methods. In *SPE Canada Unconventional Resources Conference*? SPE.
8. Slatt, R. M., & O'Brien, N. R. (2011). Pore types in the Barnett and Woodford gas shales: Contribution to understanding gas storage and migration pathways in fine-grained rocks. *AAPG Bulletin, 95*(12), 2017–2030.
9. Ramirez, W. F., & Morelli, P. T. (1981). Characterization of porous media properties and leaching capacity of spent oil shales. In *SPE Annual Technical Conference and Exhibition*? SPE.
10. Comisky, J. T., et al. (2011). Sample size effects on the application of mercury injection capillary pressure for determining the storage capacity of tight gas and oil shales. In *SPE Canada Unconventional Resources Conference*? SPE.
11. Li, H., & Wan, W. (1995). Investigation of the asphaltene precipitation process from cold lake bitumen by confocal scanning laser microscopy. In *SPE International Thermal Operations and Heavy Oil Symposium*. SPE.
12. Hampton, T. J., Querin, E. M., & McCabe, W. D. (1996). Case history: 31S C/D shale reservoirs, Monterey formation, elk hills field, California. In *SPE Western Regional Meeting*. SPE.
13. Shah, S., et al. (2013). Predicting porosity and permeability of carbonate rocks from core-scale to pore-scale using medical CT, confocal laser scanning microscopy and micro CT. In *SPE Annual Technical Conference and Exhibition*? SPE.

Chapter 3
Gas Transport Mechanisms in Nano Pores of Shale Gas Reservoirs

3.1 General Gas Flow Mechanisms

Unlike conventional gas reservoirs, gas transports in shale gas reservoirs is a complex process. Gas diffusion, viscous flow due to pressure gradient, and desorption from Kerogen coexist. Also, the gas storage forms are complex in shale reservoirs. There are three forms for the gas to store in the shale: (1) free gas stored in the fractures and pores; (2) absorbed gas stored on the surface of the bedrock; (3) dissolved gas stored in the kerogen. Gas transport in unconventional shale strata is a multi-mechanism-coupling process which is different from conventional reservoirs. In micro fractures which are inborn or induced by hydraulic stimulation, viscous flow dominates. And gas surface diffusion and gas desorption should be further considered in organic nano pores. Also, Klinkenberg effect should be considered when dealing with gas transport problem. Some scholars have studied gas transport behavior in SGRs. Bustin et al. [1] have studied effect of fabric on the gas production from shale strata. However, it is assumed that matrix does not have viscous flow and diffusion mechanisms. It is also assumed that gas transport in the fracture system obeys Darcy law which is not sufficiently accurate. Ozkan et al. [2] used a dual-mechanism approach to consider transient Darcy and diffusive flows in the matrix and stress-dependent permeability in the fractures for naturally fractured SGRs. However, they ignored effects of adsorption and desorption. Moridis et al. [3] has considered Darcy's flow, non-Darcy flow, stress-sensitive, gas slippage, non-isothermal effects, and Langmuir isotherm desorption. They found that production data from tight-sand reservoirs can be adequately represented without accounting for gas adsorption whereas there will be significant deviations if gas adsorption is omitted in SGRs. But they did not consider gas diffusion in the Kerogen. Wu and Fakcharoenphol [4] has proposed an improved methodology to simulate shale gas production, but the gas adsorption–desorption were ignored in the model. It is necessary to develop a mathematical model which incorporates all known mechanisms to describe the gas transport behavior in tight shale formations. The simulation work was greatly simplified when the apparent

C. Guo and Z. Yang, *Shale Gas Production: Concept, Models, and Techniques*, Petroleum Engineering, https://doi.org/10.1007/978-3-031-84552-9_3

permeability was first introduced by Javadpour. In 2009, the concept of apparent permeability considering Knudsen diffusion, slippage flow and advection flow was proposed Javadpour [5]. Through this method, the flux vector term can be simply expressed in the form of Darcy equation which will greatly reduce the computation complexity. Then the concept of apparent permeability was further applied in pore scale modeling for shale gas. Civan and Ziarani derived the expression for apparent permeability in the form of Knudsen number [6, 7] based on a unified Hagen-Poiseuille equation [8]. Following are those mechanisms which have been considered in simulation of shale gas reservoirs according to the literature review.

3.1.1 Darcy Flow

The advection of gas phase in porous media is commonly characterized using Darcy equation. From Darcy equation, the gas velocity is directly proportional to the gas pressure gradient. Darcy equation is applicable when the gas velocity is not high or there is no boundary shear effect. However, Darcy flow has also been considered in simulation of shale gas reservoirs. For gas flow, more commonly used equation is the extended Darcy equation which considers the Klinkenberg effect (Klinkenberg, 1941). Klinkenberg pointed that for the same rock sample and test gas, the gas permeability is different under different average pressure. Also, for the same rock sample and average pressure, the gas permeability is different if the test gas are different. Equation (25) is the general Darcy equation which considers Klinkenberg effect:

$$V = \frac{K}{\mu}\left(1 + \frac{b_k}{\overline{P}}\right)\nabla P \tag{3.1}$$

where V is the gas flow velocity, m/s; K is the Klinkenberg permeability, m^2; b_k is the Klinkenberg coefficient, MPa; μ is the gas viscosity, Pa s; $\overline{P}$ is the average gas pressure, MPa.

3.1.2 Flow in the Fracture

At high flow velocities in the near wellbore hydraulic fractures, inertia effect cannot be neglected. Due to inertia effect, the flow rate will keep constant when pressure increases. There are two reasons for this. One reason is gas deceleration within larger pores which slows down approaching gas molecules; another reason is though gas acceleration within tighter pores, gas velocity keeps constant due to viscous shearing effect [9]. So, the conventional linear Darcy's equation which assumes zero inertia effect will no longer be valid. Gas flow in hydraulic fracture follows Forchheimer flow model [3], which takes inertia effect into consideration. Forchheimer equation can be expressed as follows:

$$\frac{\partial p}{\partial x} = -\frac{\mu}{k}V + \beta p V^2 \tag{3.2}$$

where $\frac{\partial p}{\partial x}$ is the pressure gradient (Pa/m); μ is the gas viscosity (Pa s); V is the velocity (m/s); β is the Forchheimer coefficient; ρ is the fluid density (kg/m^3); k is the permeability (m^2). The first parameter on the right hand side of Eq. (26) represents viscous flow and dominates at lower velocity and the second parameter on the right hand side of Eq. (26) represents inertial forces and dominates at higher velocity where viscous forces become less dominant. Also, we can find that when β = 0, Eq. (26) changes to Darcy equation.

Many scholars have proposed methods to compute the Forchheimer β-parameter. Kutasov [10] computes the β-parameter as a function of effective permeability to gas as well as water saturation based on experiment results. Frederick and Graves [11] have developed two empirical correlations for β-parameter in two phase flow. Cooke [12] has correlated β-parameter coefficients to various proppant types.

3.1.3 Diffusion

The diffusion in porous media consists of continuum diffusion, and free molecule diffusion. Continuum diffusion can be characterized using Fick's law or Stefan-Maxwell equations [13]. Free molecule diffusion, or Knudsen diffusion occurs when the gas mean free path is close to the pore diameter, such as in the nano pore, which has been detailed in the flow mechanisms of gas flow in nano pores. In shale formation, the diffusion refers to the gas flow from high concentration zone (shale matrix system) to the low concentration zone (shale micro-fracture system) due to concentration difference. When the equilibrium achieves, the diffusion process will end. The diffusion process in shale formation can be classified into pseudo steady diffusion and unsteady diffusion. The pseudo steady diffusion can be characterized using Fick first law, which states the mole or mass flux is proportional to a diffusion coefficient times the gradient of the mole or mass concentration. The unsteady diffusion is generally characterized using Fick's second law, which states the temporal evolution of gas concentration.

3.1.4 Gas Adsorption/Desorption

Adsorption gas is one main storage form in shale gas reservoirs. The adsorption gas is about 40–50% of the total gas in place. Gas desorption is one important recovery mechanism in shale gas reservoirs. Gas transport through shale fracture network is the combination of desorption and diffusion. Cipolla et al. [14] showed that considering gas desorption will lead to improved gas recovery. Thompson et al. [15] also found that desorption gas accounts for about 17% of the expected ultimate recovery. Arogundade and Sohrabi [9] concluded that 5–15% of the total gas production are desorption gas in shale gas reservoirs.

The adsorption of gas on the surface of shale matrix is affected by temperature, pressure, gas type, and rock properties. To characterize the gas adsorption on the rock surface, there are three widely used adsorption equations: Henry equation; Freundlich equation; and Langmuir equation. Henry equation states that the adsorption volume is a linear function of pressure. With the increase of gas pressure, the adsorption volume increases. However, Henry equation assumes that the adsorption gas is ideal gas which is not valid for high pressure range. Freundlich equation states there is exponential relationship between gas adsorption volume and pressure. When the pressure is large enough, gas adsorption volume will increase much slowly. So, the pressure range for the Freundlich equation is only limited to low pressure range. The most widely used for characterizing shale gas adsorption is Langmuir isotherm equation [16]. Langmuir assumes that there is monolayer gas molecules on the surface. The pressure depletion and gas desorption are synchronized and the system can achieve equilibrium instantly. As the shale gas are mainly composed of Methane. And the temperature of shale formation is higher than the critical temperature of Methane (– 47.22 °C). So, the methane will be adsorbed on the surface as monolayer. Based on the Langmuir equation (Eq. 27), the amount of gas adsorbed on the rock surface can be computed.

$$C(P_g) = V_L \frac{P_g}{P_g + P_L} \tag{3.3}$$

where V_L is the gas Langmuir volume, scf/ton; P_L is the gas Langmuir pressure, psi; P_g is the in-situ gas pressure in the pore system, psi. Langmuir volume means the maximum amount of gas that can be adsorbed on the rock surface under infinite pressure. Langmuir pressure is the pressure when the amount of gas adsorbed is 50% of the Langmuir volume. These two parameters play important roles in the gas desorption process. For different shale gas reservoirs, the differences of Langmuir volume and Langmuir pressure lead to distinct trend of gas content. Figure 3.1 shows the adsorbed gas content (scf/ton) versus pressure (psi) based on Langmuir parameters for five major shale plays in U.S.

3.1.5 Stress, Water, and Temperature Sensitivity

For gas flow in shale formations, some scholars also consider the effect of stress, water, and temperature. Stress sensitivity refers to the change of shale permeability, porosity, total stress, effective stress, and rock properties (such as compressibility, Young's modules) with stress change. The effect of stress on matrix and fracture

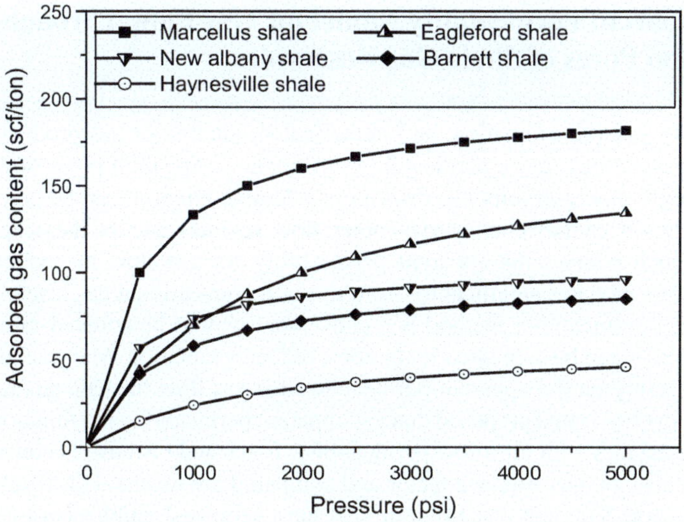

Fig. 3.1 Langmuir isotherm curves for five typical shale gas reservoirs

permeability has been studied in detail. Celis et al. [17] have simulated the oil flow in stress sensitive naturally fractured reservoirs using permeability modulus. Gutierrez et al. [18] have carried out experiments to investigate the stress dependent permeability of de-mineralized fractures in shale. Raghavan and Chin [19] have proposed a geomechanical-fluid flow model to study the pressure transient response for stress sensitive reservoirs. Franquet et al. [20] have investigated the effect of pressure dependent permeability on transient radial flow for tight gas reservoirs. Thompson et al. [15] have concluded that pressure dependent permeability is responsible for performance behavior by analyzing the well performance data of Haynesville shale. Cho et al. [21] have studied the pressure dependent natural fracture permeability in shale and its effect on shale gas well production. Pedrosa [22] studied the pressure transient responses in stress sensitive reservoirs by defining a permeability modulus.

Water sensitivity refers to the swelling of clay particles due to the existence of water. The effect can be characterized by correcting gas–water relative permeability curve or capillary pressure. Temperature permeability refers to the formation and fluid properties change due to temperature. The effect can be characterized using Peng-Robinson equation [23].

3.2 Apparent Permeability Model for Gas Flow Through Nano Pores of Shale Gas Reservoirs

Methods for reserve estimation, and numerical simulation of gas production from shale gas reservoirs require permeability information. Apparent permeability study for shale/tight gas reservoirs has drawn considerable attention as the conventional Darcy equation cannot characterize other flow regimes except the viscous flow regime, which courses the apparent permeability for gas flow in unconventional gas reservoirs deviates significantly from those in conventional gas reservoirs. And the apparent permeability method is a convenient way to be adopted in numerical simulation as it can be expressed in the form of Darcy equation. Many scholars have conducted study on the apparent permeability for gas flow in shale gas reservoirs. Klinkenberg (1941) first proposed that gas apparent permeability is a function of pressure for reservoirs with ultra-low permeability. Jones and Owens [24] have derived an empirical static gas slippage factor and compared the results with Klinkenberg's slippage factor. Sampath and Keighin [25] have proposed Klinkenberg coefficient correction for N_2 in presence of water. Ertekin et al. [26] incorporated Klinkenberg effect to consider the higher than expected gas production in shale gas reservoirs. Beskok and Karniadakis [8] proposed a unified Hagen-poiseuille-type equation to characterize gas flow in tight/shale reservoirs. The proposed model has considered multiple transport mechanisms, such as viscous flow, slip flow, transition flow, and free molecular flow. Based on the work of Beskok and Karniadakis [8], Florence et al. [27] proposed a theoretically derived apparent permeability model which is a function of Knudsen number. Freeman [28] modified the apparent permeability model of Florence to obtain a component-dependent apparent permeability. Recently, some researchers have proposed the methods to calculate the apparent permeability by coupling Darcy flow with Knudsen diffusion to describe gas flow in shale reservoirs.

3.2.1 *Beskok and Karniadakis' Model*

The unified Hagen-poiseuille-type equation which is proposed by Beskok and Karniadakis [8] can be expressed in Eq. (12). This model can consider different flow regimes based on the Knudsen number.

$$q_h = f(K_n)\frac{\pi R_h^4}{8\mu}\frac{\Delta p}{L_h} \tag{3.4}$$

$$f(K_n) = (1 + \alpha K_n)\left(1 + \frac{4K_n}{1 - bK_n}\right) \tag{3.5}$$

$$\alpha = \alpha_0 \frac{2}{\pi}\tan\left(\alpha_1 K_n^{\alpha_2}\right) \tag{3.6}$$

$$\frac{\alpha_0}{\alpha} - 1 = \frac{A}{K_n^B}, \ A > 0, \ B > 0 \tag{3.7}$$

where R_h is the pore diameter, L_h is the length of flow path, q_h is the volumetric gas flow rate through a single pipe, K_n is the Knudsen number which based on gas mean free path and pore diameter, α is the dimensionless rarefaction coefficient which can be calculated using Eq. (14) which is provided by Beskok and Karniadakis [8] or using Eq. (15) which is provided by Civan and Vafai [7]. The parameter in the b in the Eq. (13) is the empirical slip coefficient, which is equal to 1 for fully developed slip flow condition. α_0, α_1, α_2, A and B are all empirical constants which can be determined through experiments.

Equation (12) can be used to calculate the flow rate for a single pore. For a bundle of tortuous flow paths such as the membranes, the volumetric gas flow rate can be calculated using Eq. (16).

$$q = nq_h = nf(K_n)\frac{\pi R_h^4}{8\mu}\frac{\Delta p}{L_h} \tag{3.8}$$

where L_h is also the length of the flow path, n is the number of the preferential hydraulic flow paths which can be calculated using the formula provided by Civan and Vafai [7].

$$n = \frac{\phi A_b}{\pi R_h^2} \tag{3.9}$$

where ϕ is the porosity, A_b is the bulk surface area normal to the flow direction.

3.2.2 *Javadpour's Model*

Slip flow and Knudsen diffusion are two important flow mechanisms in SGRs. These two mechanisms can be combined in one equation to characterize gas flow for both conventional and unconventional SGRs. Javadpour [5] conducted a thorough study of gas flow in nano pores, deriving following formula for gas flow in nanotubes based on Knudsen diffusion and viscous force, as shown in Eqs. (18) and (19):

$$J = -\frac{1}{M_g}\left[\frac{2rM_g}{3000RT}\left(\frac{8RT}{\pi M_g}\right)^{0.5} + F\frac{r^2\rho_{avg}}{8\mu}\right]\nabla P \tag{3.10}$$

$$F = 1 + \left(\frac{8\pi RT}{M_g}\right)^{0.5}\frac{\mu_g}{P_{avg}r}\left(\frac{2}{\alpha} - 1\right) \tag{3.11}$$

The apparent permeability provided by Javadpour is expressed in Eq. (19):

$$k_{app} = \frac{2r\mu_g M_g}{3000RT\rho_{avg}}\left(\frac{8RT}{\pi M_g}\right)^{0.5} + F\frac{r^2}{8} \tag{3.12}$$

3.2.3 Florence's Model

Florence et al. [27] have derived the apparent permeability expression by approximating the apparent permeability expression of Beskok and Karniadakis [8] under slip flow condition ($\boldsymbol{\alpha} = 0, \boldsymbol{b} = -1$) for $\boldsymbol{K_n} \ll 1$, which is shown in Eq. (3.13).

$$K = K_\infty(1 + \alpha K_n)\left(1 + \frac{4K_n}{1 - bK_n}\right) \underset{slip\,flow\,condition}{\overset{\alpha=0,b=-1}{\longrightarrow}}$$
$$K_\infty\left(1 + \frac{4K_n}{1 + K_n}\right) \overset{K_n \ll 1}{\longrightarrow} K_\infty(1 + 4K_n) \tag{3.13}$$

3.2.4 Civan's Model

The apparent permeability model provided by Civan [29] is shown in Eq. (3.15), which is based on the work of Beskok and Karniadakis [8].

$$K = K_\infty f(K_n) \tag{3.14}$$

$$K_\infty = \frac{\pi R_h^2}{8\tau_h} \tag{3.15}$$

where K_∞ is the intrinsic permeability or the liquid permeability which is only related to the porous media and does not depend on the fluid types and flow conditions, τ_h is the tortuosity factor of the flow paths.

3.2.5 Sakhaee-Pour and Bryant's Model

Sakhaee-Pour and Bryant [30] also proposed a model to characterize gas apparent permeability based on Knudsen number. They regarded the method proposed by Florence et al. [27], which is one-order of Knudsen number is not accurate enough to characterize gas flow in the transition period. After fitting with the experimental data, they proposed a high-order equation of Knudsen number which is fit for the Knudsen number in the range from 0.1 to 0.8. The apparent permeability expression is shown as follows:

$$K = K_{\infty}\left(0.8453 + 5.4576K_n + 0.1633K_n^2\right) \tag{3.16}$$

3.3 Simulation Model for Gas Production from Shale Gas Reservoirs

Reservoir simulation is the most useful technology for management of subsurface oil/gas reservoirs, which can help understand shale gas reservoirs and optimize the production decisions. Recent advances in simulation and modeling technologies have enable it possible to optimize and enhance the gas production from organic shale reservoirs. Many scholars have conducted modeling of gas production from unconventional shale gas reservoirs. The key problem in simulating fractured shale gas reservoirs is how to handle the fracture matrix interaction. Many different models have been proposed to simulate fluid flow in fractured reservoirs. These methods include: (1) equivalent continuum model; (2) dual porosity model, including dual porosity single permeability model, and dual porosity dual permeability model; (3) Multiple Interaction Continua (MINC) method; (4) Multiple porosity model; (5) Discrete fracture model. The details of each model are as follows:

3.3.1 Equivalent Continuum Model

The Equivalent continuum model is first proposed by Snow [31]. Many different types of equivalent continuum models have been proposed with different assumptions, such as for isothermal flow, coupled fluid and heat flow, single phase flow, and multi-phase flow. In the equivalent continuum model, the fractured porous media is considered as a continuum media. The properties of the fracture arrays, such as direction, location, permeability, porosity, etc., are averaged into the whole porous media. This method is focused on the study of the macro flow characteristics of the porous media without considering the specific flow condition in a single fracture. This method has not been widely used in shale gas reservoir simulation as it is not easy to obtain the averaged permeability tensor and it is too conceptual. Moridis et al. [3] have constructed effective continuum shale gas reservoir simulation model with considering multi-component gas adsorption. In the effective continuum model, the shale gas reservoir is discretized and the fractures are characterized with grid cells as single planar planes or network of planar planes.

3.3.2 Dual Porosity Model

The concept of dual porosity was first proposed by Warren and Root [32]. Then, this concept was introduced into petroleum engineering by Warren and Root [32]. Dual porosity model is widely used to simulate the fluid flow in fractured porous media and many scholars have published a lot of literature based on dual porosity model [33]. There are two interacting media: the matrix blocks with high porosity and low permeability, and the fracture network with low porosity and high conductivity. According to whether there is fluid flow in the matrix, the dual porosity model can be classified into dual porosity single permeability model and the dual porosity dual permeability model.

In the dual porosity single permeability model, the flow only occurs through the fracture system and matrix are treated as the spatially distributed sinks or sources to the fracture system. In the shale gas reservoirs, the matrix system serves as the gas storage place for the free and adsorption gas. And the fracture system provides high permeability. Watson et al. [34] have performed production data analysis for Devonian shale gas reservoir using DPM and presented analytical reservoir production models for history matching and production forecasting. Bustin et al. [35] have constructed a 2D dual porosity model to simulated shale gas reservoirs which uses both experimental and field data as model input parameters. The fluid flow is governed by Darcy's law. The coupled governing equations are discretized implicitly in time and 2D space using finite difference method. Du et al. [36] simulated the hydraulic fractured shale gas reservoir as a dual porosity system. Microseismic responses, hydraulic fracturing treatments data and production history-matching analysis were applied. Ozkan et al. [2] have developed a dual-mechanism dual-porosity model to simulate the linear flow for fractured horizontal well in shale gas reservoirs.

Comparably same with dual porosity single permeability model, dual-porosity-dual-permeability model is also studied with considering the flow in the matrix additionally. The matrix properties dominate the flow from matrix to fracture. Connell and Lu have proposed an algorithm for the dual-porosity model considering Darcy flow in the fractures and diffusion in the matrix [37]. Moridis et al. [3] constructed a dual permeability model and compared it with dual porosity model and effective continuum model. Results showed that dual permeability model offered the best production performance. But they also pointed out that the deviations become more obvious during the later time of production. Renyi et al. [38] have proposed a new simulation model for shale gas reservoir based on dual porosity dual permeability model with considering multiple flow mechanisms, such as adsorption/desorption, diffusion, viscous flow, and deformation.

3.3.3 Multiple Interaction Continua Method

As an extension for dual porosity model, Pruess [39] have proposed a multiple interaction continua method (MINC) to more accurately characterize the interaction between matrix and fracture. The matrix system in every grid block for MINC method will be subdivided into a sequence of nested rings which will make it possible to more accurately calculate the interblock fluid flow. In comparison with dual porosity model, MINC method is able to describe the gradient of pressures, temperatures, or concentrations near matrix surface and inside the matrix which can provide a better numerical approximation for the transient fracture matrix interaction. Wu [40] have proposed a generalized mathematical model to simulate multiphase flow in tight fractured reservoirs using MINC method with considering following mechanisms: Klinkenberg effect, non-Newtonian behavior, non-Darcy flow, and rock deformation.

3.3.4 Multiple Porosity Model

Some scholars also proposed multiple porosity model in which matrix is further separated into two or three parts with different properties. Civan et al. [41] proposed a qual-porosity model to simulate shale gas reservoirs, including organic matter, inorganic matter, natural, and hydraulic fractures. Dehghanpour and Shirdel [42] have proposed a triple porosity model to simulate shale gas reservoirs by incorporating micro fractures into the dual porosity models. They assumed that the matrix blocks in dual porosity model is composed of sub-matrix with nano Darcy permeability, and micro fractures with milli to micro-Darcy permeability. Result of sensitivity analysis showed that the rate of wellbore pressure drop has been significantly decreased using this model. Yan et al. [43] presented a micro-scale model. In Yan's model, the shale matrix was classified into inorganic matrix and organic matrix. The organic part was further divided into two parts basing on pore size of kerogen: organic materials with vugs and organic materials with nanopores. Therefore, there are four different continua in this model: nano organic matrix, vug organic matrix, inorganic matrix, and micro fractures.

3.3.5 Discrete Fracture Model

In the discrete fracture model, the fractures are discretely modeled. The discrete fracture modeling requires unstructured gridding of the matrix fracture system using Delaunay triangulation. Each fracture is characterized as a geometrically well-defined entity. Baca et al. [44] have proposed a 2D model for solute and heat flow in fractured reservoir using DFM. Juanes et al. [45] investigated 2D and 3D single phase flow in fractured reservoir using finite element DFM. Karimi-Fard et al. [46] have presented a simplified discrete fracture model using finite volume method combined with unstructured gridding, which can be used for 2D and 3D multiphase fluid flow simulation. Gong et al. [47] has proposed a workflow to enable the DFM to be applied in reservoir scale.

In the modeling of shale gas reservoirs, gas flow from the matrix to the discrete fractures by diffusion and desorption. Mayerhofer et al. [48] have investigated the fluid flow in fractured shale gas reservoirs considering stimulated reservoir volume (SRV) using explicit fracture network [48]. Gong et al. [49] have proposed a systematic methodology of building discrete fracture model based on microseismic interpretations for shale reservoirs with considering multiple mechanisms, such as gas adsorption, non-Darcy effect, etc.

The discrete fracture model is more rigorous, however, application of this method to reservoir scale is limited due to computational intensity and lack of detailed knowledge of fracture matrix properties and spatial distribution in reservoirs. And the dual porosity model can deal with the fracture matrix interaction more easily and less computationally demanding than discrete fracture method. So, dual porosity model is the main approach for modeling fluid and heat flow in fractured porous media.

3.4 Conclusion Remark

Compared with conventional reservoirs, there are a large number of nano pores existed in the SGRs. Therefore, the gas flow mechanisms are more complex compared with conventional reservoirs. In this chapter, we presented multiple gas flow mechanisms which are existed in shale gas production process, such as viscous flow, Knudsen diffusion, slip flow, water and stress sensitivity, gas viscosity change, pore radius change, and gas adsorption/desorption. In the nano scale, the continuum flow assumption breaks down and gas molecule's collision with the walls become dominant. Apparent permeability models are widely used in the simulation of gas flow through nano pores and many scholars have conducted related research in this area. For the simulation of SGRs, many models have been proposed to simulate shale gas production, such as equivalent continuum model, dual porosity model, multiple porosity model, and the discrete fracture model. The most widely used model is dual porosity model, which contains two interacting media: the matrix blocks with high porosity and low permeability, and the fracture network with low porosity and high conductivity. The study of fluid flow mechanisms and numerical simulation of shale gas reservoirs is still in the developing period. Many models were developed to more accurately characterize gas transport behavior in the nano pores and the SGRs.

References

1. Bustin, A. M., Bustin, R. M., & Cui, X. (2008). Importance of fabric on the production of gas shales. In *SPE Unconventional Resources Conference/Gas Technology Symposium*. SPE.
2. Ozkan, E., Raghavan, R., & Apaydin, O. G. (2010). Modeling of fluid transfer from shale matrix to fracture network. In *SPE Annual Technical Conference and Exhibition*? SPE.
3. Moridis, G. J., Blasingame, T. A., & Freeman, C. M. (2010). Analysis of mechanisms of flow in fractured tight-gas and shale-gas reservoirs. In *SPE Latin America and Caribbean Petroleum Engineering Conference*. SPE.
4. Wu, Y. -S., & Fakcharoenphol, P. (2011). A unified mathematical model for unconventional reservoir simulation. In *SPE Europec featured at EAGE Conference and Exhibition*? SPE.
5. Javadpour, F. (2009). Nanopores and apparent permeability of gas flow in mudrocks (shales and siltstone). *Journal of Canadian Petroleum Technology, 48*(08), 16–21.
6. Ziarani, A. S., & Aguilera, R. (2012). Knudsen's permeability correction for tight porous media. *Transport in Porous Media, 91*, 239–260.
7. Civan, F., & Vafai, K. (2010). A review of approaches for describing gas transfer through extremely tight porous media. In *AIP Conference Proceedings*.
8. Beskok, A., & Karniadakis, G. E. (1999). Report: A model for flows in channels, pipes, and ducts at micro and nano scales. *Microscale Thermophysical Engineering, 3*(1), 43–77.
9. Arogundade, O., & Sohrabi, M. (2012). A review of recent developments and challenges in shale gas recovery. In *SPE Kingdom of Saudi Arabia Annual Technical Symposium and Exhibition*. SPE.
10. Kutasov, I. (1993). Equation predicts non-Darcy flow coefficient. *Oil and Gas Journal (United States), 91*(11).
11. Frederick Jr, D., & Graves, R. (1994). New correlations to predict non-Darcy flow coefficients at immobile and mobile water saturation. In *SPE Annual Technical Conference and Exhibition*? SPE.
12. Cooke, C., Jr. (1973). Conductivity of fracture proppants in multiple layers. *Journal of Petroleum Technology, 25*(09), 1101–1107.
13. Bird, R. B. (2002). Transport phenomena. *Applied Mechanics Reviews, 55*(1), R1–R4.
14. Cipolla, C. L., et al. (2010). Reservoir modeling in shale-gas reservoirs. *SPE Reservoir Evaluation & Engineering, 13*(04), 638–653.
15. Thompson, J. M., Nobakht, M., & Anderson, D. M. (2010). Modeling well performance data from overpressured shale gas reservoirs. In *SPE Canada Unconventional Resources Conference*? SPE.
16. Langmuir, I. (1917). The constitution and fundamental properties of solids and liquids. II. Liquids. *Journal of the American Chemical Society, 39*(9), 1848–1906.
17. Celis, V., et al. (1994). A new model for pressure transient analysis in stress sensitive naturally fractured reservoirs. *SPE Advanced Technology Series, 2*(01), 126–135.
18. Gutierrez, M., Øino, L., & Nygaard, R. (2000). Stress-dependent permeability of a de-mineralised fracture in shale. *Marine and Petroleum Geology, 17*(8), 895–907.
19. Raghavan, R., & Chin, L. Y. (2002). Productivity changes in reservoirs with stress-dependent permeability. In *SPE Annual Technical Conference and Exhibition*? SPE.
20. Franquet, M., et al. (2004). Effect of pressure-dependent permeability in tight gas reservoirs, transient radial flow. In *PETSOC Canadian International Petroleum Conference*. PETSOC.
21. Cho, Y., Apaydin, O. G., & Ozkan, E. (2013). Pressure-dependent natural-fracture permeability in shale and its effect on shale-gas well production. *SPE Reservoir Evaluation & Engineering, 16*(02), 216–228.
22. Pedrosa Jr, O. A. (1986). *Pressure transient response in stress-sensitive formations*. In *SPE Western Regional Meeting*. SPE.
23. Peng, D.-Y., & Robinson, D. B. (1976). A new two-constant equation of state. *Industrial & Engineering Chemistry Fundamentals, 15*(1), 59–64.
24. Jones, F. O., & Owens, W. (1980). A laboratory study of low-permeability gas sands. *Journal of Petroleum Technology, 32*(09), 1631–1640.

25. Sampath, K., & Keighin, C. W. (1982). Factors affecting gas slippage in tight sandstones of cretaceous age in the Uinta basin. *Journal of Petroleum Technology, 34*(11), 2715–2720.
26. Ertekin, T., King, G. R., & Schwerer, F. C. (1986). Dynamic gas slippage: A unique dual-mechanism approach to the flow of gas in tight formations. *SPE Formation Evaluation, 1*(01), 43–52.
27. Florence, F. A., et al. (2007). Improved permeability prediction relations for low-permeability sands. In *SPE Rocky Mountain Petroleum Technology Conference/Low-Permeability Reservoirs Symposium*. SPE.
28. Freeman, C. (2010). A numerical study of microscale flow behavior in tight gas and shale gas. In *SPE Annual Technical Conference and Exhibition*. Society of Petroleum Engineers.
29. Civan, F. (2010). Effective correlation of apparent gas permeability in tight porous media. *Transport in Porous Media, 82*, 375–384.
30. Sakhaee-Pour, A., & Bryant, S. L. (2012). Gas permeability of shale. *SPE Reservoir Evaluation & Engineering, 15*(04), 401–409.
31. Snow, D. T. (1968). Rock fracture spacings, openings, and porosities. *Journal of the Soil Mechanics and Foundations Division, 94*(1), 73–91.
32. Warren, J. E., & Root, P. J. (1963). The behavior of naturally fractured reservoirs. *Society of Petroleum Engineers Journal, 3*(03), 245–255.
33. Kazemi, H. (1969). Pressure transient analysis of naturally fractured reservoirs with uniform fracture distribution. *Society of Petroleum Engineers Journal, 9*(04), 451–462.
34. Watson, A. T., et al. (1989). *An analytical model for history matching naturally fractured reservoir production data*. In *SPE Oklahoma City Oil and Gas Symposium/Production and Operations Symposium*. SPE.
35. Bustin, R. M., et al. (2008). Impact of shale properties on pore structure and storage characteristics. In *SPE Shale Gas Production Conference*. SPE.
36. Du, C. M., et al. (2010). Modeling hydraulic fracturing induced fracture networks in shale gas reservoirs as a dual porosity system. In *SPE International Oil and Gas Conference and Exhibition in China*. SPE.
37. Connell, L. D., & Lu, M. (2007). A dual-porosity model for gas reservoir flow incorporating adsorption behaviour—Part II. Numerical algorithm and example applications. *Transport in Porous Media, 69*(2), 139–158.
38. Renyi, C., Linsong, C., & Peiqing, L. (2017). Mathematical model of shale gas reservoirs coupled by desorption, stress sensitivity and fractures finite conductivity. *Physical and Numerical Simulation of Geotechnical Engineering, 26*, 27.
39. Pruess, K. (1992). *Brief guide to the MINC-method for modeling flow and transport in fractured media*.
40. Wu, Y. -S., et al. (2009). A multi-continuum model for gas production in tight fractured reservoirs. In *SPE Hydraulic Fracturing Technology Conference and Exhibition*. SPE.
41. Civan, F., Rai, C. S., & Sondergeld, C. H. (2011). Shale-gas permeability and diffusivity inferred by improved formulation of relevant retention and transport mechanisms. *Transport in Porous Media, 86*, 925–944.
42. Dehghanpour, H., & Shirdel, M. (2011). A triple porosity model for shale gas reservoirs. In *SPE Canada Unconventional Resources Conference*? SPE.
43. Yan, B., Wang, Y., & Killough, J. E. (2016). Beyond dual-porosity modeling for the simulation of complex flow mechanisms in shale reservoirs. *Computational Geosciences, 20*, 69–91.
44. Baca, R., Arnett, R., & Langford, D. (1984). Modelling fluid flow in fractured-porous rock masses by finite-element techniques. *International Journal for Numerical Methods in Fluids, 4*(4), 337–348.
45. Juanes, R., Samper, J., & Molinero, J. (2002). A general and efficient formulation of fractures and boundary conditions in the finite element method. *International Journal for Numerical Methods in Engineering, 54*(12), 1751–1774.
46. Karimi-Fard, M., Durlofsky, L. J., & Aziz, K. (2004). An efficient discrete-fracture model applicable for general-purpose reservoir simulators. *SPE Journal, 9*(02), 227–236.

47. Gong, B., Karimi-Fard, M., & Durlofsky, L. J. (2008). Upscaling discrete fracture characterizations to dual-porosity, dual-permeability models for efficient simulation of flow with strong gravitational effects. *SPE Journal, 13*(01), 58–67.
48. Mayerhofer, M. J., et al. (2006). Integration of microseismic fracture mapping results with numerical fracture network production modeling in the Barnett shale. In *SPE Annual Technical Conference and Exhibition*? SPE.
49. Gong, B., et al. (2011). Detailed modeling of the complex fracture network of shale gas reservoirs. In *SPE Middle East Unconventional Resources Conference and Exhibition*. SPE.

Chapter 4
Single Component Gas Transport Model in Shale Gas Nanopores

Based on the flow mechanism of shale gas in nanopores which is presented in this chapter, scholars have put forward different gas transport models. Javadpour studied the migration of shale gas by atomic mechanical microscopy and proposed an apparent permeability model considering the effect of Knudsen diffusion and gas slippage on the gas transport in nanopores of tight rock layers [1]. Cheng [2] considered multiple transport mechanisms and derived a simplified model of apparent permeability [2]. Xia et al. considered the nonlinear and nonequilibrium gas adsorption and surface diffusion, and introduced the transport equation of shale gas in dual porosity media, which synthesizes all kinds of mechanisms [3]. Su et al. constructed gas transport equations for inorganic and organic shale respectively [4]. Guo et al. considered the effect of gas desorption on gas viscosity, and derived the weak form of flow equation in dual porosity media [5]. Liu et al. [6] built a mathematical model considering the surface diffusion of adsorbed shale gas, and discussed the effects of energy heterogeneity, isosteric heat of adsorption, and non-isothermal desorption on the diffusion of gas molecule [6].

For shale reservoirs, the existence of water distribution in the organic pores is still a controversial topic, but most scholars have a consensus on the distribution of water in the inorganic pores [7]. Haghighi et al. found that under natural conditions, the actual shale reservoirs often have a certain degree of water saturation [8]. Rose et al. and Gasparika et al. tested a large number of shale samples in Canada, the United States and Europe, and found that shale adsorption capacity decreased under water- existing conditions [9, 10]. Zhang et al. found that the thickness of the water film arranged on the surface of shale clay minerals is about 0.4 nm [9]. Spears et al. found that the volume of irreducible water in inorganic shale clay could reach 7.2% of shale sample volume [12]. This shows that the water distribution in the pores of inorganic shale cannot be ignored. Through molecular simulation experiments, Jin et al. studied the interaction between methane and water molecules on the pore surface of clay [13]. Based on thermodynamics, Li et al. established a quantitative model of water film thickness and discussed the distribution characteristics of water

C. Guo and Z. Yang, *Shale Gas Production: Concept, Models, and Techniques*, Petroleum Engineering, https://doi.org/10.1007/978-3-031-84552-9_4

saturation [14]. Although they have studied the water distribution characteristics of inorganic shale clearly, there is still little discussion on the influence of water characteristics on gas transport in nanopores.

4.1 Three-Phase Adsorption and Desorption Mechanism

In shale gas reservoirs, methane usually exists in shale pores in three forms: adsorbed gas on the shale matrix and pore surface, free gas in the pores and natural fractures, and dissolved gas in water and kerogen. The adsorbed gas is the main component, which accounts for about 85% of the total gas content in shale gas reservoirs.

For the adsorption model of shale gas on the pore surface of shale, scholars generally adopt the Langmuir isotherm adsorption model, which is given as follows:

$$V_1 = V_L \frac{P}{P + P_L} \tag{4.1}$$

In the equation, V_1 is the gas adsorption volume, the unit is m^3/kg; V_L is the saturated adsorption volume of Langmuir, the unit is m^3/kg; P_L is the pressure constant of Langmuir when the gas adsorption is 50%, the unit is Pa.

According to the research of many scholars, there is a layer of water film on the surface of minerals in inorganic shale due to the existence of clay minerals. The existence of water film can impose an impact on the adsorption and desorption of shale gas. Li et al. considered the existence of water film on the inorganic pore surface [15], and proposed following equation:

$$V = \beta \frac{V_H P}{P + P_H} + (1 - \beta) \frac{V_L P}{P + P_L} \tag{4.2}$$

In the equation, V is the gas adsorption volume, the unit is m^3/kg; V_L is the solid–gas saturated adsorption volume (Langmuir saturated adsorption volume), the unit is m^3/kg; P_L is the solid–gas pressure constant (Langmuir pressure constant), the unit is Pa; V_H is the liquid–gas saturated adsorption capacity, and the unit is m^3/kg; P_H is the liquid–gas pressure constant, the unit is Pa. β is defined as the water molecular coverage ratio, dimensionless, which is the ratio of wetting area of water molecules to total pore area, dimensionless. The β can be calculated using following equation:

$$\beta = \frac{A_H}{A_{total}} \tag{4.3}$$

Furthermore, there is a relationship between the water molecule coverage ratio and irreducible water saturation. By combining the definition of water molecular coverage with the capillary irreducible water saturation equation, and then obtained the equation that correlates irreducible water saturation with water molecular coverage ratio, which is shown as follows:

$$S_w = \left[1 - \left(1 - \frac{h}{r}\right)^2\right] \cdot \beta \tag{4.4}$$

In the equation, S_w is the irreducible water saturation in the capillary, dimensionless; r is the pore radius; h is the thickness of water molecular layer.

Considering the three-phase adsorption model, the equation of the total gas adsorption in the presence of water film can be obtained, which can be given by the following equation:

$$q_a = \frac{\rho_s MV}{V_m} = \frac{\rho_s M}{V_m}\left[\beta\frac{V_H P}{P + P_H} + (1 - \beta)\frac{V_L P}{P + P_L}\right] \tag{4.5}$$

In the equation, ρ_s is the rock density, the unit is kg/m^3; M the gas mole mass, the unit is kg/mol; V_m is the gas molar volume under the standard condition (0 °C, 1 atm), the unit is m^3/mol; q_a is gas adsorption capacity, the unit is kg/m^3.

Considering the surface coverage of the adsorbed gas, the surface coverage ratio can be obtained from the solid–gas adsorption model, which is as follows:

$$\theta_1 = \frac{V_1}{V_L} = \frac{P}{P + P_L} \tag{4.6}$$

Based on the liquid–gas adsorption model, the surface coverage ratio can be obtained as follows:

$$\theta_2 = \frac{V}{V_H} = \frac{P}{P + P_H} \tag{4.7}$$

In the three-phase adsorption model, the total surface coverage ratio of the adsorbed gas is as follows:

$$\theta = \beta \cdot \frac{P}{P + P_H} + (1 - \beta) \cdot \frac{P}{P + P_L} \tag{4.8}$$

When the adsorbed gas desorbs from the surface of nanopores in the shale matrix, the gas molecules will fall off from the pore surface, which will increase the pore radius. In inorganic clay shale, the gas desorption process considering the three-phase adsorption model is shown in Fig. 4.1. Thus, the pore radius can be modified as follows:

$$r_e = r_{\max} - d_H \cdot \beta - d_C \cdot \theta = r_{\max} - d_H \cdot \beta - d_C \cdot \left[\beta\frac{P}{P + P_H} + (1 - \beta)\frac{P}{P + P_L}\right] \tag{4.9}$$

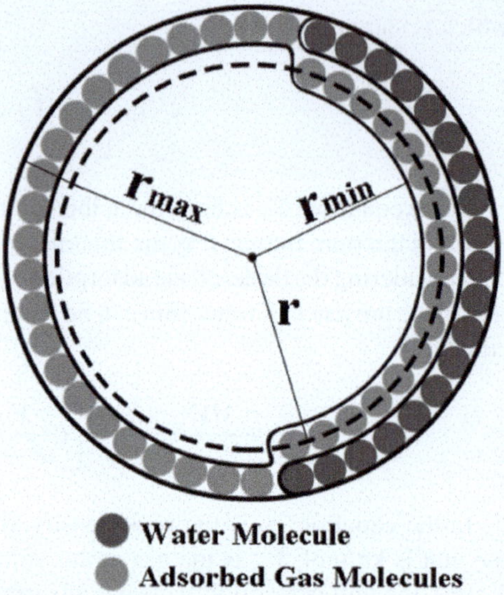

Fig. 4.1 Effect of three-phase desorption on pore radius. $r_{\max}$ is the pore radius without any fillers; r_{min} is the pore radius filled with water molecules and gas molecules; r is the equivalent pore radius

Where, r_e is the equivalent pore radius, d_H is the water molecular layer thickness, d_C is the methane molecular diameter, $r_{\max}$ is the pore radius without any adsorption gas in the inorganic pore, which is also called the ideal maximum pore radius.

According to Li's research, the water molecular layer thickness is related to the relative humidity of the gas and pore radius [15]. In this chapter, we consider the flow of shale gas with the relative humidity extremely low. In this case, the water molecular layer thickness is almost the thickness of one layer of water molecule.

The desorption of gas can lead to the change of porosity. Thus, the porosity also needs to be modified, which is shown as follows:

$$\phi = \chi \cdot \phi_{\max} \frac{r_e^2}{r_{\max}^2} = \chi \cdot \phi_{\max} \frac{\left\{ r_{\max} - d_H \cdot \beta - d \cdot \left[\beta \frac{P}{P+P_H} + (1-\beta) \frac{P}{P+P_L} \right] \right\}^2}{r_{\max}^2} \tag{4.10}$$

In the equation, $\phi_{\max}$ is the porosity under ideal maximum pore radius, which is called the ideal maximum porosity, ϕ is the intrinsic porosity, and χ is the comprehensive coefficient, which is related to formation heterogeneity, and it is defined as the ratio of actual measured porosity to theoretical porosity at the same pore radius.

Considering the three-phase adsorption model, we modified the pore radius and porosity. It can be found that these parameters are not only related to the pressure, but also related to the existence of water molecules on the water-wet inorganic pore surface.

4.2 Single Component Gas Transport Model in Shale Nanopores

Under the initial reservoir condition, the shale gas is mainly adsorbed in the shale matrix. With the pressure of shale gas reservoir decreases, the shale gas gradually desorbs. The increasing gas concentration difference will be produced. In this process, the gas molecules will diffuse through the nanopores of the matrix, as shown in Fig. 4.2. By making reasonable assumptions and simplifying the complex underground geological conditions, we constructed a mathematical model to describe shale gas flow in nanopores.

4.2.1 Model Assumptions

(1) The gas adsorption obeys the three-phase adsorption model of methane, water film and shale clay, and it is the monomolecular adsorption.
(2) The process of gas flow is isothermal.
(3) The shale matrix is slightly compressible.
(4) The effect of gravity can be ignored.
(5) Only shale gas is involved in flow, the relative humidity of the gas is extremely low.

4.2.2 Gas Transport Model in Nanopores

The equation of continuity of shale gas in nanopores is given as follows:

$$\frac{\partial \rho_{total}}{\partial t} + \nabla(\rho \cdot \mathbf{v}) = q \tag{4.11}$$

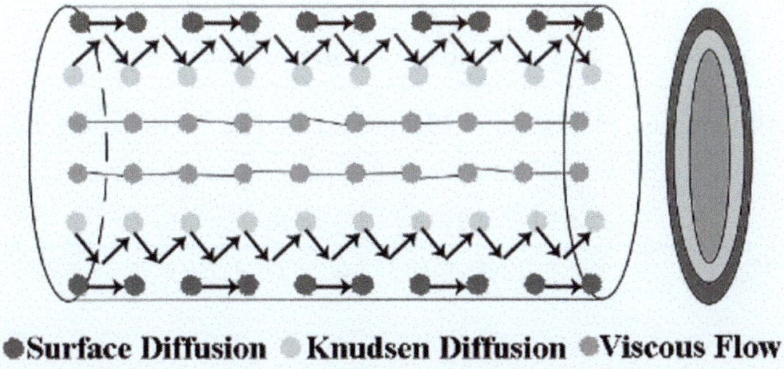

Fig. 4.2 Flow mechanisms of shale gas in nanopores. Red solid dots represent surface diffusion; yellow ones represent Knudsen diffusion; green ones represent viscous flow

where, ρ_{total} is the total amount of gas in nanopores, the unit is kg/m^3; $\mathbf{v}$ is the gas flow velocity (vector form), the unit is m/s; q is the source term.4

4.2.3 Time-Gradient Changes of Shale Gas Density

There are both adsorbed gas and free gas in the matrix. In the presence of water film, the adsorbed gas is given by Eq. (4.4), while the free gas is given by the real gas equation of state, which is shown as follows:

$$\rho = \frac{PM}{ZRT} \tag{4.12}$$

In the equation, ρ is the gas density and the unit kg/m^3; R is the ideal gas constant, the value is 8.314 J/(mol K); Z is the compressibility factor, dimensionless; T is the temperature, the unit is K.

Then the change of the total gas density with time is as follows:

$$\frac{\partial \rho_{total}}{\partial t} = \frac{\partial(\phi\rho)}{\partial t} + \frac{\partial\left[(1-\phi)q_a\right]}{\partial t} \tag{4.13}$$

It can be seen from the above formula that the change of the total gas density consists of two parts, the free gas and the adsorbed gas, which are discussed separately as follows.

The time-gradient change of the free gas is further simplified as follows:

$$\frac{\partial(\phi\rho)}{\partial t} = \phi\frac{\partial \rho}{\partial t} + \rho\frac{\partial \phi}{\partial t} = \left(\frac{\phi\rho}{P} + \rho\frac{\partial \phi}{\partial P}\right)\frac{\partial P}{\partial t} \tag{4.14}$$

Similarly, the time-gradient change of the adsorbed gas is approximately as follows:

$$\frac{\partial\left[(1-\phi)q_a\right]}{\partial t} = \left[(1-\phi)\frac{\partial q_a}{\partial P} - q_a\frac{\partial \phi}{\partial P}\right]\frac{\partial P}{\partial t} \tag{4.15}$$

By substituting Eqs. (4.14) and (4.15) into Eq. (4.13), the following equation can be obtained:

$$\frac{\partial \rho_{total}}{\partial t} = \left\{\frac{\phi\rho}{P} + (1-\phi)\frac{\rho_s M}{V_m}\left[\frac{(1-\beta)\cdot V_L\cdot P_L}{(P+P_L)^2} + \frac{\beta\cdot V_H\cdot P_H}{(P+P_H)^2}\right] + (\rho - q_a)\frac{\partial \phi}{\partial P}\right\}\frac{\partial P}{\partial t} \tag{4.16}$$

In the above equation, the term $(\rho - q_a)\frac{\partial \phi}{\partial P}$ can be ignored (The reason why the term $\left(\rho - q_g\right)\frac{\partial \phi}{\partial P}$ can be ignored is presented in the Appendix B). Thus, the equation can be approximated as follows:

$$\frac{\partial \rho_{total}}{\partial t} \approx \rho \left\{ \frac{\phi}{P} + (1-\phi)\frac{\rho_s M}{\rho V_m}\left[\frac{(1-\beta)\cdot V_L \cdot P_L}{(P+P_L)^2} + \frac{\beta \cdot V_H \cdot P_H}{(P+P_H)^2} \right] \right\} \frac{\partial P}{\partial t} \tag{4.17}$$

In order to simplify the solution equation, the water storage coefficient S, is defined as the follows:

$$S = \frac{\phi}{P} + (1-\phi)\frac{\rho_s M}{\rho V_m}\left[\frac{(1-\beta)\cdot V_L \cdot P_L}{(P+P_L)^2} + \frac{\beta \cdot V_H \cdot P_H}{(P+P_H)^2} \right] \tag{4.18}$$

The unit of storage coefficient S is 1/Pa. Then Eq. (4.16) can be simplified as:

$$\frac{\partial \rho_{total}}{\partial t} = \rho S \frac{\partial P}{\partial t} \tag{4.19}$$

4.2.4 Gas Velocity Term in Nanopores

The flow of gas in nanopores mainly includes Knudsen diffusion, slippage effect, surface diffusion, and etc. In this chapter, we mainly consider the equation of gas motion under the effect of Knudsen diffusion and gas slippage effect, as follows:

$$\mathbf{v} = \mathbf{v}_k + \mathbf{v}_d \tag{4.20}$$

In the equation, $\mathbf{v}$ is the velocity of gas flow (vector form), the unit is m/s; $\mathbf{v}_k$ is the velocity of Knudsen diffusion (vector form), the unit is m/s; $\mathbf{v}_d$ is velocity of viscous flow considering slippage effect (vector form), the unit is m/s.

Because the thickness of the water film is very small, here we ignore the influence of the water film on the Knudsen diffusion and the viscous flow.

In the equation, the Knudsen diffusion velocity is given as follows:

$$\mathbf{v}_k = -\frac{2rM}{3RT\rho}\sqrt{\frac{8RT}{\pi M}}\nabla P \tag{4.21}$$

The motion of the viscous flow in the long straight circular capillary follows the Hagen-Poiseuille equation, and the following equation is obtained:

$$\mathbf{v}_0 = -\frac{r^2}{8\mu}\nabla P \tag{4.22}$$

In the equation, $\mathbf{v}_0$ is the velocity of viscous flow without considering the effect of gas slippage, and μ is the gas viscosity, the unit is Pa s. In the nanopores, the average free path of the gas molecule is comparable to the pore size, so the gas will slip on the pore surface (Klinkenberg, 1941). Brown et al. introduced the Klinkenberg effect coefficient F to modify the viscous flow in the capillary tube [16], and the equation is as follows:

$$F = 1 + \sqrt{\frac{8\pi RT}{M}}\frac{\mu}{\overline{P}r}\left(\frac{2}{\alpha} - 1\right) \tag{4.23}$$

In the equation, α is the tangential momentum correction coefficient, dimensionless; $\overline{P}$ is the average pressure, the unit is Pa. Then, the velocity of viscous flow considering the Klinkenberg effect is as follows:

$$\mathbf{v}_d = F \cdot \mathbf{v}_0 = \left[1 + \sqrt{\frac{8\pi RT}{M}}\frac{\mu}{\overline{P}r}\left(\frac{2}{\alpha} - 1\right)\right]\frac{r^2}{8\mu}\nabla P \tag{4.24}$$

Combine Eqs. (4.20), (4.23) and (4.9), we can obtain the following equation:

$$\mathbf{v} = -\frac{\frac{2\mu r_e M}{3RT\rho}\sqrt{\frac{8RT}{\pi M}} + \left[1 + \sqrt{\frac{8\pi RT}{M}}\frac{\mu}{\overline{P}r_e}\left(\frac{2}{\alpha} - 1\right)\right]\frac{r_e^2}{8}}{\mu}\nabla P \tag{4.25}$$

It can be found that the Eq. (4.25) is similar to Darcy flow equation. The apparent permeability k_{app} can be obtained:

$$k_{app} = \frac{2\mu r_e M}{3RT\rho}\sqrt{\frac{8RT}{\pi M}} + \left[1 + \sqrt{\frac{8\pi RT}{M}}\frac{\mu}{\overline{P}r_e}\left(\frac{2}{\alpha} - 1\right)\right]\frac{r_e^2}{8} \tag{4.26}$$

In the equation, k_{app} is called apparent permeability, it is not only related to the intrinsic properties of reservoirs, but also related to fluid properties, formation temperature, formation pressure and other factors. Further considering the difference between apparent permeability and reservoir intrinsic permeability, the ratio of apparent permeability to reservoir intrinsic permeability can be obtained:

$$\frac{k_{app}}{k_d} = \frac{16\mu M}{3r_e RT\rho}\sqrt{\frac{8RT}{\pi M}} + 1 + \sqrt{\frac{8\pi RT}{M}}\frac{\mu}{\overline{P}r_e}\left(\frac{2}{\alpha} - 1\right) \tag{4.27}$$

Then, the equation of gas motion after the introduction of apparent permeability can be simplified to the form similar to Darcy's law, as follows:

$$\vec{v} = -\frac{k_{app}}{\mu}\nabla P \tag{4.28}$$

4.2.5 The Mathematical Equation of Shale Gas in Nanopores

Combining Eqs. (4.11), (4.19) and (4.28), the flow equation of shale gas matrix can be obtained as follows:

$$\rho S\frac{\partial P}{\partial t} - \nabla \cdot \left(\frac{\rho \cdot k_{app}}{\mu}\nabla P\right) = q \tag{4.29}$$

In the equation, S is the water storage coefficient, k_{app} is the apparent gas permeability, ϕ is the porosity, and r_e is the effective pore radius. The four proposed parameters can be calculated as follows:

$$S = \frac{\phi}{P} + (1-\phi)\frac{\rho_s M}{\rho V_m}\left[\frac{(1-\beta)\cdot V_L \cdot P_L}{(P+P_L)^2} + \frac{\beta \cdot V_H \cdot P_H}{(P+P_H)^2}\right] \tag{4.30}$$

$$k_{app} = \frac{2\mu r_e M}{3RT\rho}\sqrt{\frac{8RT}{\pi M}} + \left[1 + \sqrt{\frac{8\pi RT}{M}}\frac{\mu}{\overline{P}r_e}\left(\frac{2}{\alpha} - 1\right)\right]\frac{r_e^2}{8} \tag{4.31}$$

$$\phi = \chi \cdot \phi_{\max}\frac{\left\{r_{\max} - d_H \cdot \beta - d_C \cdot \left[\beta\frac{P}{P+P_H} + (1-\beta)\frac{P}{P+P_L}\right]\right\}^2}{r_{\max}^2} \tag{4.32}$$

$$r_e = r_{\max} - d_H \cdot \beta - d_C \cdot \left[\beta\frac{P}{P+P_H} + (1-\beta)\frac{P}{P+P_L}\right] \tag{4.33}$$

4.3 Model Solution

Generally speaking, shale gas transport equation in nanopores is a system of the nonlinear partial differential equation with variable coefficients. The analytical solution is very difficult and sometimes impossible to obtain. Therefore, we often use numerical analysis method to discretize the equations and obtain the numerical solution. One of these numerical methods is called the finite element method. In this chapter, we use the finite element method to solve Eq. (4.29) and analyze the effect of water film and the significant pore size range.

The basic parameters required for solving Eq. (4.29) and values of some parameters are shown in the Tables 4.1 and 4.2.

We build a shale gas reservoir of 100 m × 100 m, where point A is a point in the shale gas reservoir, the coordinate is (10, 6), which will be analyzed later, and the marked boundary is a horizontal well. The model is shown in Fig. 4.3. Because of the small reservoir area, we do not consider the heterogeneity of the reservoir. Also, we think that there are all inorganic pores in this small area of the reservoir.

The initial and boundary conditions for the model are as follows:

Table 4.1 Basic parameter expression

Parameter name	Expression
Three-phase adsorption volume	$V = \beta \frac{V_H P}{P+P_H} + (1-\beta)\frac{V_L P}{P+P_L}$
Water molecular coverage ratio	$\beta = \frac{A_H}{A_{total}}$
Gas adsorption capacity	$q_a = \frac{\rho_s M}{V_m}\left[\beta \frac{V_H P}{P+P_H} + (1-\beta)\frac{V_L P}{P+P_L}\right]$
Pore radius	$r_e = r_{\max} - d_H \cdot \beta - d_C \cdot \left[\beta \frac{P}{P+P_H} + (1-\beta)\frac{P}{P+P_L}\right]$
Porosity	$\phi = \chi \cdot \phi_{\max} \frac{\left\{r_{\max} - d_H \cdot \beta - d_C \cdot \left[\beta \frac{P}{P+P_H} + (1-\beta)\frac{P}{P+P_L}\right]\right\}^2}{r_{\max}^2}$
Water storage coefficient	$S = \frac{\phi}{P} + (1-\phi)\frac{\rho_s M}{\rho V_m}\left[\frac{(1-\beta)\cdot V_L \cdot P_L}{(P+P_L)^2} + \frac{\beta \cdot V_H \cdot P_H}{(P+P_H)^2}\right]$
Apparent permeability	$k_{app} = \frac{2\mu r_e M}{3RT\rho}\sqrt{\frac{8RT}{\pi M}} + \left[1 + \sqrt{\frac{8\pi RT}{M}}\frac{\mu}{Pr_e}\left(\frac{2}{\alpha} - 1\right)\right]\frac{r_e^2}{8}$
Ratio of apparent permeability to rock absolute permeability	$\frac{k_{app}}{k_d} = \frac{16\mu M}{3r_e RT\rho}\sqrt{\frac{8RT}{\pi M}} + 1 + \sqrt{\frac{8\pi RT}{M}}\frac{\mu}{Pr_e}\left(\frac{2}{\alpha} - 1\right)$

Table 4.2 Value of basic parameters

Parameter name	Values and units	The meaning of parameters
$P1$	3 MPa	Pressure in the horizontal well
$P2$	50 MPa	Initial pressure of shale gas reservoir
μ	2.22e−5 Pa s	Gas viscosity
$r_{\max}$	3 nm	Ideal maximum pore radius
d_C	0.38 nm	Gas molecular mean diameter
d_H	0.4 nm	Water molecular layer thickness
P_H	44.56 MPa	Gas–liquid adsorption pressure constant
P_L	5 MPa	Langmuir pressure constant
V_L	0.003 m^3/kg	Langmuir saturated adsorption volume
V_H	0.003 m^3/kg	Gas–liquid saturated adsorption volume
R	8.314 J/(mol K)	Ideal gas constant
M	0.039948 kg/mol	Gas molar mass
ρ_s	2600 kg/m^3	Density of shale rock
α	0.8	Slip effect tangential momentum correction factor
$\phi_{\max}$	0.1	Ideal maximum porosity
V_m	22.414 L/mol	Gas molar volume under standard condition
Z	1	Gas compressibility factor
T	300 K	The temperature of the stratum
D	2 m	Diameter of horizontal well
χ	1	Comprehensive coefficient

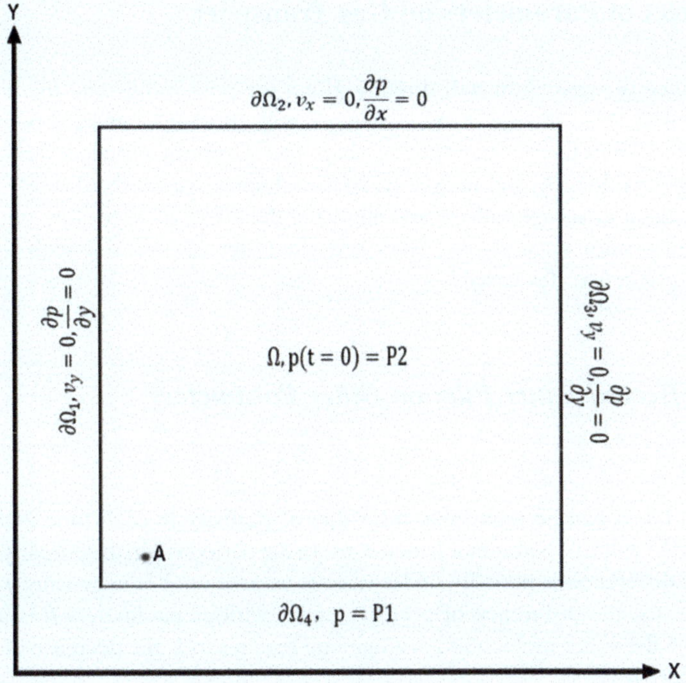

Fig. 4.3 Geometrical model in this chapter: a shale gas reservoir of 100 m × 100 m

$$\partial\Omega_1, v_y = 0, \frac{\partial P}{\partial y} = 0; \tag{4.34}$$

$$\partial\Omega_2, v_x = 0, \frac{\partial P}{\partial x} = 0; \tag{4.35}$$

$$\partial\Omega_3, v_y = 0, \frac{\partial P}{\partial y} = 0; \tag{4.36}$$

$$\partial\Omega_4, P = P_1; \tag{4.37}$$

$$\Omega, P(t = 0) = P_2. \tag{4.38}$$

4.4 Effect of Parameters on Gas Transport

The existence of water film will directly affect the pore radius and the amount of gas adsorption. Also, the pore radius change will affect the apparent permeability and porosity. Therefore, water film in shale has a certain influence on shale gas flow parameters directly and indirectly. In this chapter, we set the discrete variable, water molecular coverage ratio β, and analyzed the effect of water film on porosity, pressure, adsorption capacity etc. And then, the range of pore size with significant water film effect was discussed.

4.4.1 Effect of Water Film on Other Parameters

Porosity

From Fig. 4.4, it can be found that the value of porosity in each time is gradually reduced as the value of parameter β increases. In the period when the porosity changes slowly, the difference of porosity under various parameters β is increasing with time. At the t = 0 s, the difference of porosity under various parameters β is also very large. When the water molecular coverage rate reaches to 1, the decrease of porosity can be about 18.1%.

When the ideal maximum pore radius is fixed, the pressure in the shale decreases gradually with time, which results in desorption of the adsorbed gas. Thus, the pore radius and porosity will increase. Since the porosity is proportional to the square of the radius, with the desorption of the gas, the difference of the pore radius at the same time with the presence of the water film will be larger and larger. At the initial time, when $\beta = 1$, the water film completely covers the pore surface, and the gas adsorption reaches its maximum at this time, which can be considered as the double layer adsorption. When β decreases, the adsorption gradually becomes the single layer solid–gas adsorption model.

Gas reservoir pressure

The curves of pressure varying with time at each β and the cloud diagram of gas reservoir pressure distribution with β equal to 0 and 1 at $t = 2 \times 10^7$ s are as shown in Figs. 4.5 and 4.6 respectively. From Fig. 4.5, we can know that the pressure distribution is as follows: the closer to the edge of the horizontal well, the pressure changes more quickly. The farther away from the horizontal well, the less obvious the pressure change. Compared with the left and right figure, the pressure drop is faster when the parameter β is larger in the same space position. As shown in Fig. 4.6, each pressure curve decreases sharply at first, then decreases slowly. As the value of β increases, the value of each pressure decreases gradually.

During the period of sharp change of pressure, pressure difference (Δp) is the main influencing factor and the influence caused by other factors can be ignored. In the period of slow change of pressure, pressure difference (Δp) is no longer the main

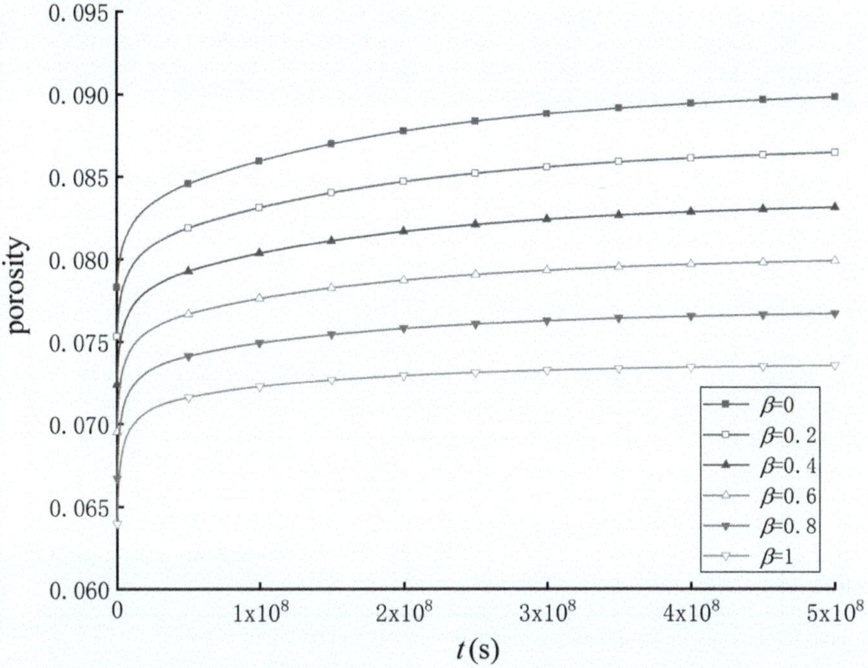

Fig. 4.4 Effect of water film on porosity variation with time. As the value of parameter β increases, the value of porosity in each time is gradually reduced

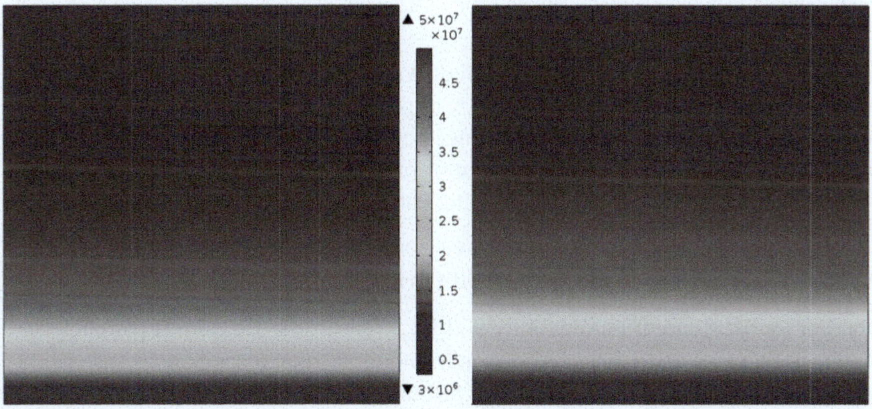

Fig. 4.5 The pressure distribution map of the gas reservoir when $t = 2 \times 10^7$ s, the unit is Pa. The left: $\beta = 0$; the right: $\beta = 1$

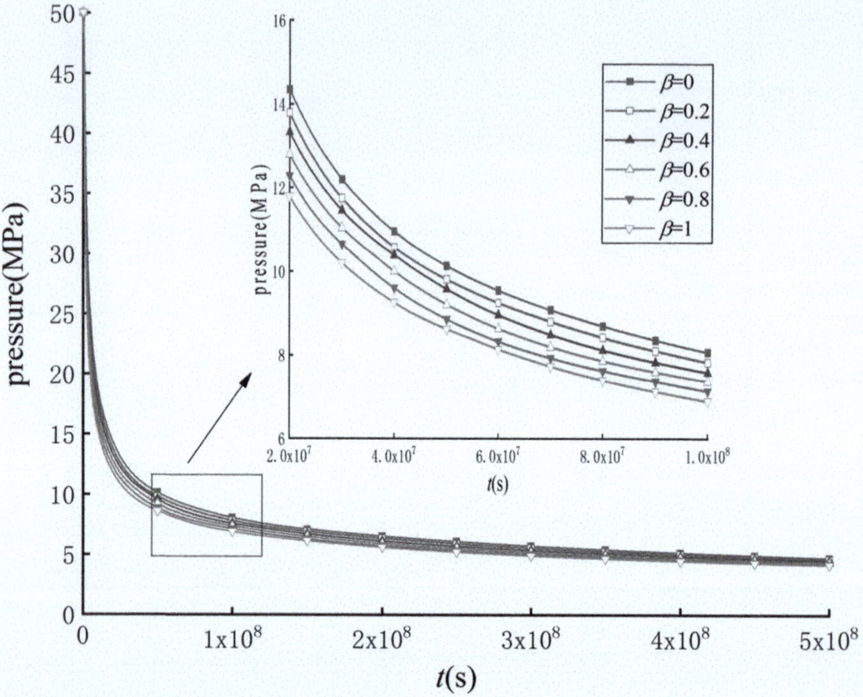

Fig. 4.6 Effect of water film on pressure variation with time. As the value of parameter β increases, the value of pressure in each time is gradually reduced

factor and other factors cannot be ignored. Water film affects apparent permeability of gas flow and also influences the change of pressure, which results in the difference of pressure at different β. At the same time, the value of each pressure curve decreases with the increase of β. Also, the change of pressure in turn will affect the changes of physical parameters When the pressure changes rapidly, the physical quantity of flow also changes sharply.

Apparent permeability and permeability ratio kapp/kd

The curves of apparent permeability and k_{app}/k_d vs. time for each β parameter are shown in Figs. 4.7 and 4.8. It can be seen from Fig. 4.7 that the apparent permeability increases with time. With the increase of β, the apparent permeability curve decreases. And with time going, the apparent permeability difference becomes more and more obvious. As shown in Fig. 4.8, with the increase of β, the ratio of apparent permeability to intrinsic permeability increases and becomes more and more obvious with time going. When the water molecular coverage rate reaches to 1, the decrease of apparent permeability is 8.4%, and k_{app}/k_d will be 9.96%.

With the decrease of the reservoir pressure, the apparent permeability and k_{app}/k_d will increase. The apparent permeability is also related to the pore radius: the larger the pore radius is, the larger the apparent permeability is. However, k_{app}/k_d

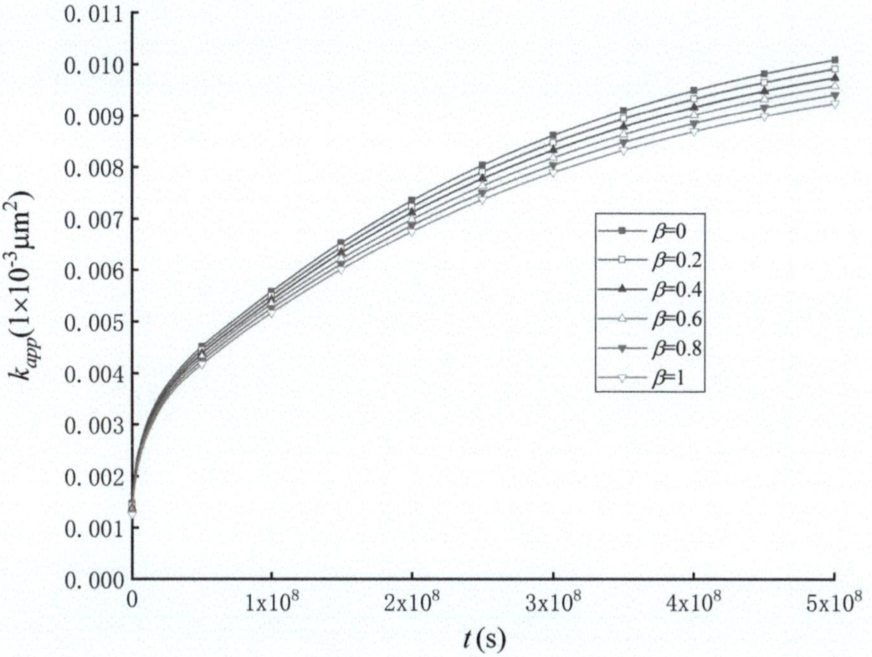

Fig. 4.7 Effect of water film on apparent permeability curve variation with time. As the value of parameter β increases, the value of apparent permeability curve in each time is gradually reduced

decreases with the increase of pore radius. Due to the existence of water film, the pore radius becomes small. And the larger the value of β is, the higher the water film coverage is, and the more obvious the decreasing of pore radius is. With time going, the influence of water film is becoming greater. However, it can be found that the pressure values of gas reservoirs under each β are almost the same and the difference is very small. So, the apparent permeability and k_{app}/k_d are mainly related to the change of β. From the expression of apparent permeability and k_{app}/k_d, we can know that k_{app}/k_d increases with the increase of β. With the passage of time, the difference becomes larger. While the apparent permeability decreases with the increase of β, and the difference becomes larger with time.

Gas adsorption capacity

The gas adsorption capacity under each β is shown in Fig. 4.9. From Fig. 4.9, it can be found that each adsorption capacity curve decreases sharply at first, then decreases slowly to the minimum. With the increase of β, the adsorption amount of each adsorption capacity curve at each time gradually decreased. Also, it can be seen that the larger the β value is, the faster the adsorption capacity curve decreases overall in the period from 0 to 2.5×10^7 s.

When the pressure and the maximum adsorption capacity are fixed, the larger the adsorption pressure constant is, the smaller the gas adsorption capacity is. This

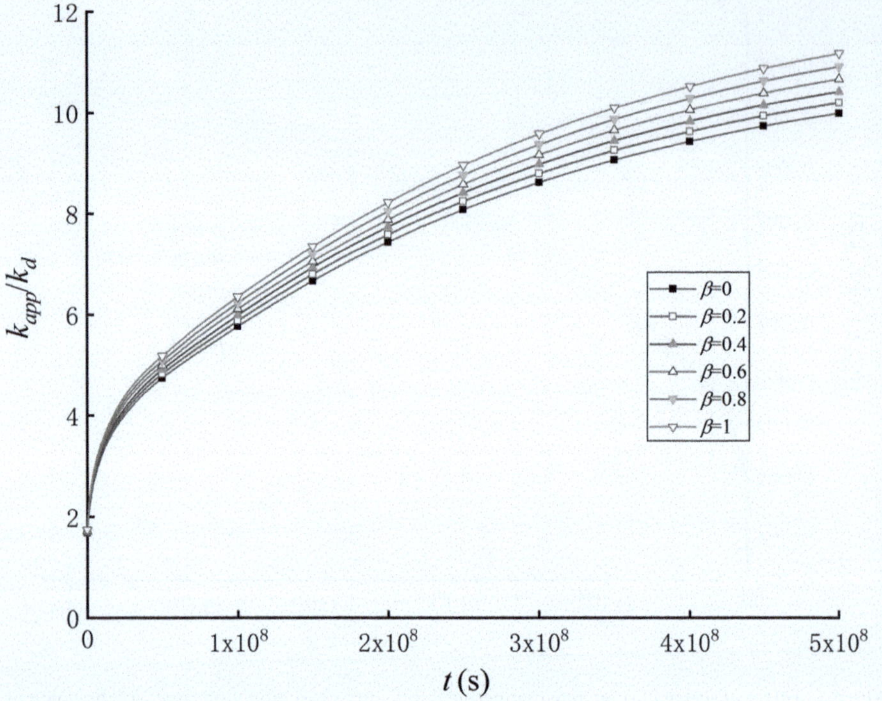

Fig. 4.8 Effect of water film on k_{app}/k_d curve variation with time. As the value of parameter β increases, the value of k_{app}/k_d curve in each time gradually increases

reflects the gas adsorption ability of different substances. By comparing the data, it is known that the adsorption ability of water molecules to gas is smaller than the adsorption ability of the pore surface to gas. Therefore, under the same pressure conditions, the larger the water film coverage is, the lower the adsorption ability of shale is, and the smaller the gas adsorption capacity is. At the same time, gas molecules are more easily desorbed from the surface of water membrane when the pressure drops rapidly, which makes the adsorption capacity decrease more rapidly.

Shale gas production rate

For the shale gas production rate, consider the situation of perfect wells, we can calculate it by integral the area fraction of the flow velocity along the semi-cross-sectional area of the horizontal well. The diameter of the horizontal well is D. The definition is as follows:

$$q_t = \iint vdA = \int \frac{D\pi}{2} vdl = \int \frac{D\pi}{2} vdx \tag{4.39}$$

In order to further discuss the effect of water film on shale gas production, we introduce the cumulative gas production parameter, which is defined as the integral

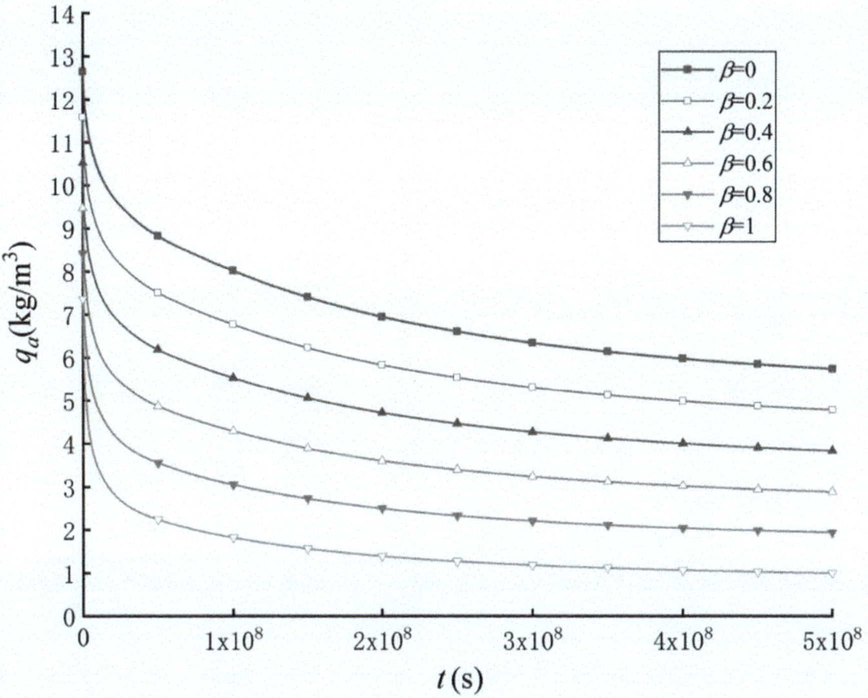

Fig. 4.9 Effect of water film on gas adsorption capacity curve variation with time. As the value of parameter β increases, the value of gas adsorption capacity curve in each time is gradually reduced

of shale gas production rate to time. The definition is as follows:

$$Q = \int q_t dt = \iint \frac{D\pi}{2} v dx dt \tag{4.40}$$

By means of numerical integration, we obtained Figs. 4.10 and 4.11. In Fig. 4.10, it can be found that the gas production rate curves decrease sharply initially and then decrease slowly. In the sub-figure of Fig. 4.10, with the increase of β, the value of gas production rate decreases gradually. We can find that the initial values of gas production rates decrease gradually with the increase of β. In total, the gas production rate is not sensitive to the change of water film parameter. However, in Fig. 4.11, we can see that with the increase of β, the cumulative gas production decreases obviously. And when the water molecular coverage rate reaches to 1, the accumulative production will decrease about 49.6%.

The flow velocity is related to the apparent permeability and the pressure-gradient. The pressure-gradient at each β value is approximately the same, so the percolation velocity is mainly related to the apparent permeability. When the value of β increases, the apparent permeability decreases gradually. Thus, the flow velocity decreases. So the integral with the same upper and lower limit of velocity will also be decreased.

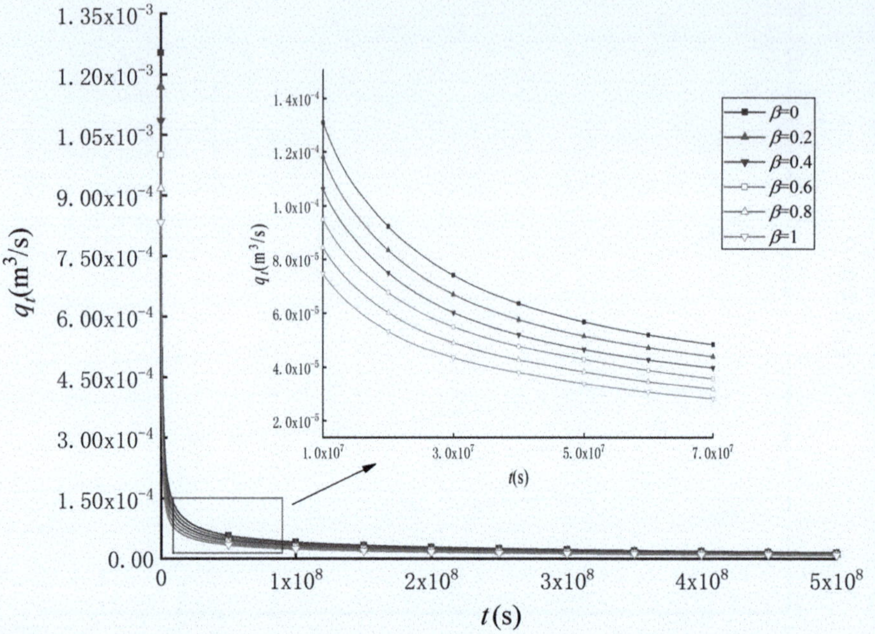

Fig. 4.10 Effect of water film on gas production rate curve variation with time. As the value of parameter β increases, the value of gas production rate in each time is gradually reduced

At the time t = 0 s, the apparent permeability decreases with the increase of β, so the initial gas production rate decreases with the increase of β. The change of gas production rate determines the change of cumulative gas production, which makes the cumulative gas production decrease with the increase of β. The results show that the existence of water film reduces the cumulative gas production.

4.4.2 *Discussion on the Range of Pore Size with Significant Water Film Effect*

In this section, we studied the effect of water film on permeability ratio under the conditions of different $r_{\max}$ and β. After scanning different $r_{\max}$ and β, Fig. 4.12 can be obtained. In Fig. 4.12, it can be found that the smaller pore size is, the more significant the water film effect is. As $r_{\max}$ increases, the change of k_{app}/k_d with β is not obvious. In the range of $r_{\max}$ less than or equal to 11 nm, the change of k_{app}/k_d with β is significant. But the range is only the perceptual understanding. So, we need to carry the quantitative study for the range of pore size with the significant water film effect.

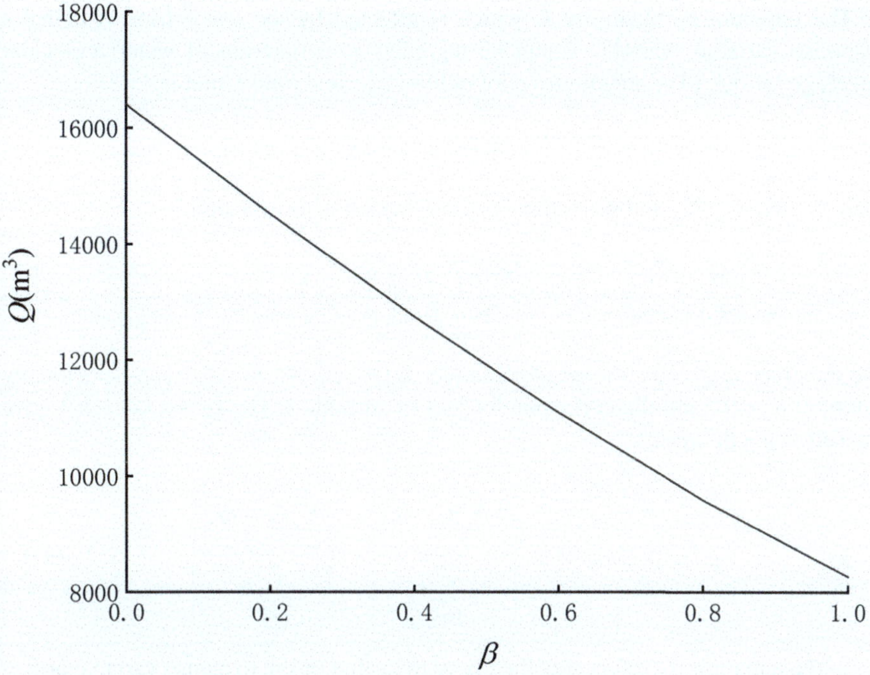

Fig. 4.11 Relationship between cumulative gas production and β

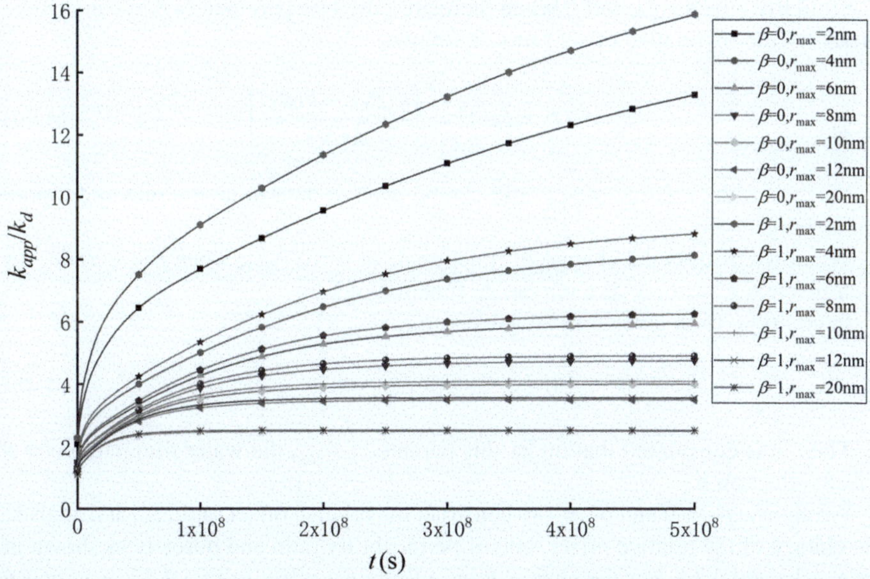

Fig. 4.12 The effect of water film under different pore radius. The effect is not obvious with the increase of pore radius

The apparent permeability k, which is affected by the water film, is called a response variable. Variable k can be regarded as a function of water molecular coverage ratio β, ideal maximum pore radius $r_{\max}$, and time t, that is:

$$k = k(\beta, r_{\max}, t) \tag{4.41}$$

Here we use k_{app}/k_d and ϕ to characterize k, which is as followed:

$$k = \frac{16\mu M}{3r_e RT\rho}\sqrt{\frac{8RT}{\pi M}} + 1 + \sqrt{\frac{8\pi RT}{M}}\frac{\mu}{\overline{P}r_e}\left(\frac{2}{\alpha} - 1\right) \text{ or } k = \phi_{\max}\frac{r_e^2}{r_{\max}^2} \tag{4.42}$$

For the given $r_{\max}$ and t, the set of parameter β, $\{\beta_0, \beta_1, \beta_2, \beta_3 \ldots \beta_n\}$ (β_0 specifically refers to $\beta = 0$), and the corresponding set of variable k, $\{k_0, k_1, k_2, k_3 \ldots k_n\}$, give the following definition:

$$\delta_n(\beta, r_{\max}, t) = \left|\frac{\frac{1}{n}\sum_{i=1}^{n} k_i - k_0}{\frac{1}{n}\sum_{i=1}^{n} k_i}\right| = \left|\frac{\sum_{i=1}^{n} k_i - nk_0}{\sum_{i=1}^{n} k_i}\right| \tag{4.43}$$

In the equation, k_0 refers specifically to the value of the response variable at $\beta = 0$. For the given $r_{\max}$ and t, $\delta_n(\beta, r_{\max}, t)$ is called the absolute offset ratio for the given $r_{\max}$ and t. δ_n is used to quantitatively characterize the effect of water film.

Similarly, given $r_{\max}$ and t, when there are only two parameters β, it can also be characterized by relative offset ratio, as follows:

$$\delta^2 = \delta^2(\beta, r_{\max}, t) = \left|\frac{k(\beta_1) - k(\beta_2)}{k(\beta_1)}\right| \tag{4.44}$$

Moreover, the relative offset ratio can be used to determine the maximum pore radius range.

When we discuss the change of function δ_n or δ^2 to $r_{\max}$, when β is certain, for each given $r_{\max}$, δ_n or δ^2 is a univariate function of t.

In addition, for a given critical value ε, if

$$\delta_{n\max} \le \varepsilon, \ or\ \delta_{\max}^2 \le \varepsilon \tag{4.45}$$

Then, it is considered that under this parameter $r_{\max}$, the water film parameter β has no effect on k.

We choose $\beta_2 = 0$ and $\beta_1 = 1$ to construct the relative offset ratio, δ^2, and analyze the change of the relative offset ratio of permeability ratio and porosity as shown in Figs. 4.13 and 4.14. For the change trend of relative offset in Fig. 4.13, we choose the critical value $\varepsilon = 0.02$. As shown in Fig. 4.13, we can see that the smaller pore size is, the larger the ratio is, and we can also know that the smaller pore size is, the

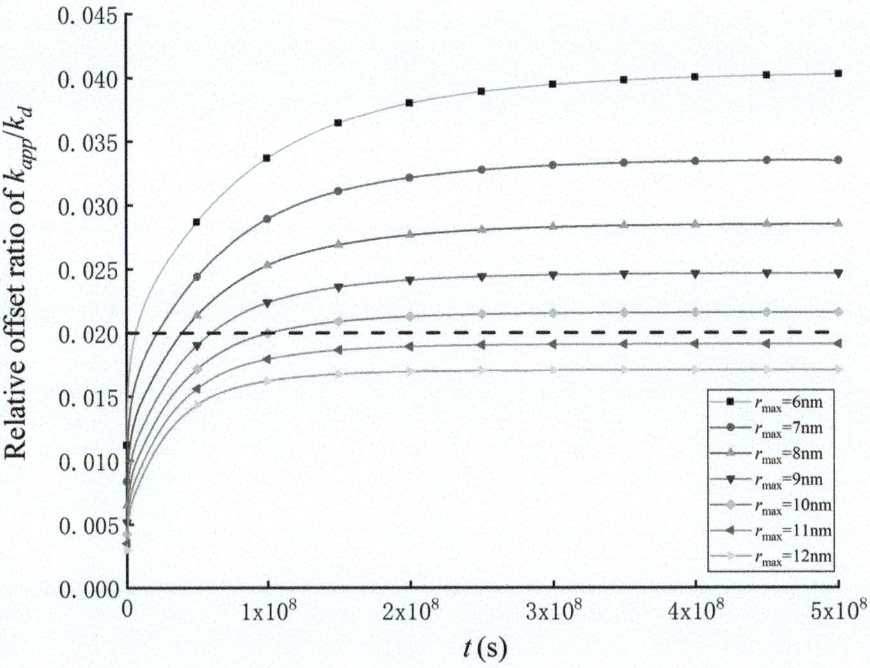

Fig. 4.13 Relative offset ratio of porosity curve variation with time

more significant the effect is. When $r_{\max} = 11\text{nm}$, $\delta^2_{\max} < 0.02$, it can be considered that the water film has no effect on k_{app}/k_d at this time. And in Fig. 4.14, when $r_{\max} = 29$ nm, it can be considered that the water film has no effect on ϕ at this time. It shows that the effect of water film on gas flow can be ignored for the pore radius above 11 nm because of the weak effect of water film. When the pore radius is beyond 29 nm, the effect of water film on reservoir physical property can be ignored. At the same time, it reflects the complexity of gas flow mechanism in nanopores.

4.5 Remark

In this chapter, we introduced the gas transport model for only one component (i.e. methane) in shale nanopores. By deriving the model and model solving, following conclusion have been drawn:

(1) By introducing the mechanism of three-phase adsorption, the equation of effective pore radius and effective porosity were obtained. Then, a new transport model of shale gas which has universality with considering three-phase adsorption was constructed;

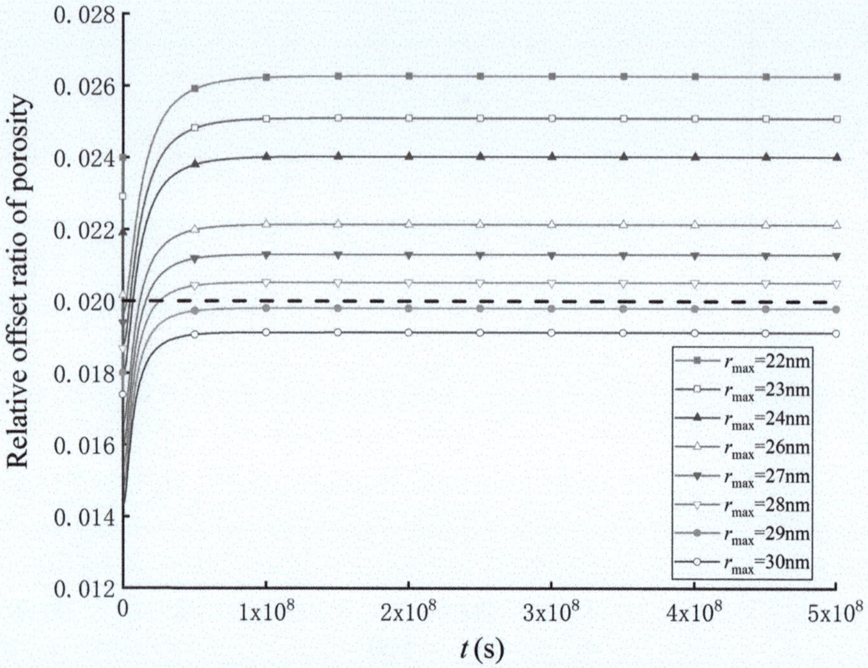

Fig. 4.14 Relative offset ratio of k_{app}/k_d curve variation with time

(2) The existence of water film on shale pore surface has a negative effect on porosity, apparent permeability and gas adsorption. That is to say, the existence of water film reduces the value of each parameter and makes the reservoir physical property and gas flow property of shale gas reservoir become worse;
(3) The existence of shale gas water film also has an influence on the actual gas production rate and cumulative gas production. The simulation results in this chapter show that water film can reduce gas production and even reduces the economic benefits of the gas reservoir in actual production;
(4) In a small pore size range, the effect of water film is significant. And the smaller pore size is, the more significant the effect is. Based on the critical value ε given in this chapter, for gas flow in nanopores, when r_{max} is below 11 nm, the effect of pore water film is obvious. And for reservoir physical property, it can be ignored when r_{max} is beyond 29 nm. The influence of water film can be neglected in large pores.

References

1. Javadpour, F. (2009). Nanopores and apparent permeability of gas flow in mudrocks (shales and siltstone). *Journal of Canadian Petroleum Technology, 48*(08), 16–21.
2. Cheng, Y. (2012). Impact of water dynamics in fractures on the performance of hydraulically fractured wells in gas-shale reservoirs. *Journal of Canadian Petroleum Technology, 51*(02), 143–151.
3. Guo, C., Wei, M., & Liu, H. (2015). Modeling of gas production from shale reservoirs considering multiple transport mechanisms. *PLoS ONE, 10*(12), e0143649.
4. Liu, X., et al. (2021). Shale gas transport in nanopores: Contribution of different transport mechanisms and influencing factors. *Energy & Fuels, 35*(3), 2033–2047.
5. Hu, Y., et al. (2015). Microscopic dynamics of water and hydrocarbon in shale-kerogen pores of potentially mixed wettability. *SPE Journal, 20*(01), 112–124.
6. Ahmad, M., & Haghighi, M. (2013) Water saturation evaluation of Murteree and Roseneath shale gas reservoirs, Cooper Basin, Australia using wire-line logs, focused ion beam milling and scanning electron microscopy. In *SPE Asia Pacific Unconventional Resources Conference and Exhibition*. SPE.
7. Ross, D. J., & Bustin, R. M. (2009). The importance of shale composition and pore structure upon gas storage potential of shale gas reservoirs. *Marine and Petroleum Geology, 26*(6), 916–927.
8. Gasparik, M., et al. (2014). Geological controls on the methane storage capacity in organic-rich shales. *International Journal of Coal Geology, 123*, 34–51.
9. Zhang, C., Shan, W., & Wang, X. (2019). Quantitative evaluation of organic porosity and inorganic porosity in shale gas reservoirs using logging data. *Energy Sources, Part A: Recovery, Utilization, and Environmental Effects, 41*(7), 811–828.
10. Spears, R. W., et al. (2011) Shale gas core analysis: strategies for normalizing between laboratories and a clear need for standard materials. In *SPWLA Annual Logging Symposium*. SPWLA.
11. Jin, Z., & Firoozabadi, A. (2014). Effect of water on methane and carbon dioxide sorption in clay minerals by Monte Carlo simulations. *Fluid Phase Equilibria, 382*, 10–20.
12. Li, J., et al. (2016). Water distribution characteristic and effect on methane adsorption capacity in shale clay. *International Journal of Coal Geology, 159*, 135–154.
13. Li, J., et al. (2015). Methane adsorption model under pore gas liquid solid three-phase interaction in shale clay. *Journal of China Coal Society, 40*(7), 1580–1587.
14. Brown, G. P., et al. (1946). The flow of gases in pipes at low pressures. *Journal of Applied Physics, 17*(10), 802–813.
15. Xia, Y., Jin, Y., Chen, M., et al., (2015). Mathematical model of shale gas percolation. Sci Bull 60, 2259–2271
16. Su, Y., Sheng, G., Wang, W., et al., (2016). Multi-media coupled flow model for shale gas reservoir. Nat. Gas. Ind. 36, 52–59

Chapter 5
Two Component Gas Transport Mechanisms in Shale Gas Nanopores

Injection carbon dioxide to enhance shale gas recovery (CO_2-ESGR) is an effective method to improve gas production and also realize the CO_2 geo-sequestration. In the CO_2-ESGR system, there are two component gases (TCG) co-existed. The gas transport and adsorption mechanisms are different from the single component gas system. However, most of current work is still focused on single component gas and many mechanisms have not been considered in the study of CO_2-ESGR study. In order to conduct reliable numerical simulation for CO_2-ESGR, clear understanding of the transport and adsorption mechanisms for TCG system is necessary. In this chapter, we focused on two key issues regarding CO_2-ESGR: the fluid transport and competitive adsorption. The fundamental mechanisms and models for pure gas and TCG transport have been comprehensively summarized. The adsorption models for TCG system have been reviewed and analyzed, such as extended Langmuir model, ideal adsorption solution model, and lattice density functional model, etc. Also, the factors affecting the competitive adsorption of TCG in shale have been illustrated, including organic matter content, shale composition, pore structure, water content, and pressure, etc. The problems existed in current research of transport and adsorption models for TCG have been analyzed. The issues which are necessary to be considered in CO_2-ESGR technology for more accurate numerical simulation and more reliable production prediction have been proposed. In this endeavor, an overview of the relevant mechanisms and models for the study of CO_2-ESGR has been presented to reflect the current state of knowledge and to suggest the direction for future research.

For the study of two component gases (TCG) diffusion, rare work has been conducted in SGRs. However, some scholars have carried out research on gas diffusion for TCG in coalbed methane (CBM) reservoirs, which can provide some foundations or references for the study in SGRs. Wei et al. have performed numerical study for enhanced CBM recovery by using CO_2 injection based on the Maxwell–Stefan diffusion theory [1]. The effect of gas concentration on diffusion has been analyzed in their study. Fathi et al. have constructed the one-dimensional triple porosity-dual permeability model for CH_4-CO_2 TCG system [2]. The improved Maxwell–Stefan

C. Guo and Z. Yang, *Shale Gas Production: Concept, Models, and Techniques*, Petroleum Engineering, https://doi.org/10.1007/978-3-031-84552-9_5

model was proposed to characterize TCG transport in micropores (0.8–2 nm) and macropores (> 50 nm). Results show that surface diffusion is an important factor which leads to competitive adsorption. Sun et al. have simulated CBM development with CO_2 injection based on Maxwell–Stefan dual-diffusion model and the results have provided some references for the study of CO_2-ESGR [3].

On the whole, the numerical study of CO_2-ESGR is still insufficient and cannot follow the development of fluid flow study in nano scale. Commercial numerical simulation software cannot characterize the gas transport mechanisms accurately and many latest mechanisms have not been effectively considered in current study. And also, the TCG diffusion mechanisms have not been clearly understood. Most studies still follow Fick's diffusion law, which cannot characterize the bidirectional diffusion in TCG system. As the foundation for accurate numerical simulation of CO_2-ESGR, the study of TCG transport mechanisms is still required urgently but corresponding models are not sufficient. In this chapter, the related mechanisms of TCG transport in shale and competitive adsorption have been presented. Models of TCG competitive adsorption have been reviewed and factors affecting the competitive adsorption of TCG in shale have been illustrated. And also, the problems existed in current research and models have been analyzed.

5.1 Fundamental Mechanisms for Two-Component Gas Transport in SGRs

5.1.1 Adsorption Mechanism

There are mainly three forms for shale gas to be stored in shale gas reservoirs: adsorption gas on the surface of shale nano pores (can up to 85% of the total gas content), free gas in the micro-fractures or nano pores, and dissolved gas in organic matters or kerogen. Efficient production of adsorption gas is the key to achieve high recovery in SGRs. Adsorption is the phenomenon which occurs at the two-phase interface of a substance. Specifically, for the shale gas adsorption, the two-phase interface refers to the contact surface between the shale gas and the pore wall. In the two-phase interface, some component molecules are enriched and others are displaced, which is called adsorption. The adsorption mechanism can be divided into physical adsorption and chemical adsorption. Kang et al. found that the adsorption of methane on shale is mainly physical adsorption [4]. When gas is adsorbed on shale, shale has the ability to absorb, which is called adsorbent. The gas is the material adsorbed on the shale, which is called adsorbate. Brunauer, Deming [5] found that the force between adsorbent and adsorbent is weak van der Waals, which is often called physical adsorption and is reversible. Choi et al. found that in the physical adsorption process, van der Waals force is accompanied by electrostatic force, which can accommodate more than one single molecular layer. Physical adsorption is an exothermic process, and the range of adsorption heat is 8–40 kJ/mol [6].

5.1.2 Gas Migration Mechanism

The pores of shale gas reservoirs (SGRs) are composed by various types of nanoscale pores. Due to the pore confinement effect, the gas transport in nano pore is no longer continuous flow and the conventional Darcy equation is invalid. Therefore, the migration mechanisms of carbon dioxide and methane in SGRs mainly include: viscous flow, Knudsen diffusion, surface diffusion, and molecular diffusion, etc. Surface diffusion is caused by adsorption gas decomposition, and molecular diffusion occurs between gas molecules of different components. The flow pattern of gas in the microporous medium mainly depends on the relative magnitude of gas mean molecule free path and physical properties of the porous medium. Javadpour et al. [7] found that the flow of gas in SGRs is a complex process with multiple migration mechanisms coexisted. The flow pattern of gas in SGRs from microscopic to macroscopic scales has been proposed. The division of gas flow regimes in SGRs can be characterized by Kundsen number. According to the *Kn* number, the flow regimes in the SGR can be classified into four categories. When the *Kn* is less than 0.01, the flow regime can be characterized using the conventional Darcy flow or viscous flow; when $0.01 < Kn < 0.1$, the flow is slip flow ($0.01 < Kn < 0.1$) and the slip effect or Klinkenberg effect should be considered; when $0.1 < Kn < 10$, the flow is transitional flow and the Darcy's law with Knuden's correction can be applied; when $Kn > 10$, the flow is free molecular flow, the direct simulation monte carlo (DSMC) method or the Lattice Boltzmann equation can be used. For different flow regimes as listed in Table 5.1, different fluid transport mechanisms and corresponding mathematical characterization models should be applied.

Table 5.1 Knudsen number and flow regime classification

Flow regime	Range of Knudsen number	Model to be applied
Viscous or continuous flow	$Kn < 0.01$	Darcy's equation
Slip flow	$0.01 < Kn < 0.1$	Darcy's equation with slip effect correction
Transition flow	$0.1 < Kn < 10$	Darcy's law with Knuden's correction
Free molecular flow	$Kn > 10$	Direct simulation monte carlo (DSMC) method or the Lattice Boltzmann Equation

5.2 Two-Component Gas Transport Mechanisms and Related Models

For more reliable simulation of the CO_2-ESGR which includes TCG transport in shale nano pores, not only the mechanisms for the pure gas transport should be considered (such as viscous flow, surface diffusion, and Knudsen diffusion, etc.), but also the mechanisms due to interaction between TCG should be taken into account (such as molecular diffusion, bi-directional diffusion, and competitive adsorption, etc.). Many studies have been conducted for one component gas transport in SGRs and different approaches have been proposed to simulate gas transport in nano pores, such as Burnett equation method [8], Lattice-Boltzmann (LB) method [9], molecular dynamics (MD) method [10], and direct simulation monte carlo (DSMC) method [11], etc. Also, many analytical models for gas transport in nano pores of SGR have been gradually proposed to simplify the calculation and simulation burden. In this chapter, the fundamental transport mechanisms which should be considered in future CO_2-ESGR studies are summarized.

5.2.1 *Viscous Flow*

If the pore diameter of shale is much larger than the mean free path of shale gas molecules, the gas transport is mainly driven by pressure difference. This migration mode is called viscous flow. According to the Hagen-Poiseuille equation, the mass flux due to viscous flow can be expressed as follows:

$$J_{vs} = -\rho \frac{r^2}{8} \frac{1}{\mu} \nabla p \tag{5.1}$$

where, J_{vs} is the mass flux due to viscous flow, kg/(m^2·s); ρ is the density of shale gas, kg/m^3; μ is the viscosity of shale gas, Pa s; p is the pressure, Pa; r is the pore radius, m.

5.2.2 *Knudsen Diffusion*

If the pore diameter of shale is comparable to the mean free path of shale gas molecules, the gas migration is mainly caused by the collision of shale gas molecules with the wall. This migration mode is called the Knudsen diffusion. The Knudsen diffusion equation for the actual gas is:

$$J_k = -MD_k \nabla C \tag{5.2}$$

$$D_k = \frac{2r}{3}\sqrt{\frac{8ZRT}{\pi M}} \tag{5.3}$$

where, J_k is the mass flux due to Knudsen diffusion, kg/(m^2 s); D_k is Knudsen diffusion coefficient, m^2/s; M is molar mass, kg/mol; Z is gas compression coefficient, dimensionless; R is the gas constant, 8.314 J/(mol K); C represents the molar concentration of methane molecules, mol/m^3.

5.2.3 Surface Diffusion

In shale gas reservoirs (SGRs), the volume of adsorbed gas usually accounts for 20–85% of the total pore volume, so there is surface diffusion of adsorbed gas in the micro-nano pores of SGRs. The mass flow equation caused by surface diffusion is calculated by Fick's law, which is shown as follows [12]:

$$J_s = -MD_s\nabla C_s \tag{5.4}$$

where, J_s is the mass flux due to surface diffusion, kg/(m^2·s); D_s is the surface diffusion coefficient of shale gas, m^2/s; C_s is the adsorption capacity of shale gas, mol/m^3.

5.2.4 Molecular Diffusion

When there are two component gases (TCG) in the shale pores, the gas molecules of different components have different molecular velocities, and the gas molecules of different components undergo energy transfer due to concentration gradient. This migration mode is called the molecular diffusion. The mass flow equation due to molecular diffusion can be expressed as follows:

$$J_a = -D_{AB}\rho\nabla w_A \tag{5.5}$$

$$D_{AB} = 1.883\times 10^{-2}T^{\frac{3}{2}}\sqrt{0.001\left(\frac{1}{M_A}+\frac{1}{M_B}\right)}\cdot\frac{1}{p\sigma_{AB}\Omega_{AB}} \tag{5.6}$$

where, J_a is the mass flux due to molecular diffusion, kg/(m^2 s); D_{AB} is the molecular diffusion coefficient, m^2/s; w_A is the mass fraction of component A; σ_{AB} is the collision potential diameter of gas molecules (m); Ω_{AB} is the dimensionless molecular constant; T is the temperature, (K); M_A and M_B are the mole mass of component A and B, (kg/mol).

5.2.5 Bidirectional Diffusion

For the TCG transport, the bidirectional diffusion is also an important mechanism. However, this mechanism is rarely considered in current numerical simulation of CO_2-ESGR. Lee et al. specifically deduced the calculation of bidirectional diffusion by determining the forward and reverse solute fluxes. $J_S^{forward}$ and $J_S^{reverse}$ can be calculated by the following mass balance equation.

$$C_D^{final} V_D^{inital} + J_w A_m t - C_D^{initial} V_D^{intal} = J_S^{forward} A_m t \tag{5.7}$$

$$C_F^{final} V_F^{intal} - J_w A_m t - C_F^{intial} V_F^{initial} = J_S^{reverse} A_m t \tag{5.8}$$

where, $J_S^{forward}$ is forward solute fluxes; $J_S^{reverse}$ is reverse solute fluxes; J_w is the water flux; A_m is the membrane area; t is the measured time; C is concentration; V is volume; and the subscripts D and F represent draw and feed solutions, respectively.

5.3 Gas Transport Models in Nano Pores

For the gas transport in shale nano pores, conventional Darcy equation under continuum flow is no longer applied due to the effect of nano scale. Besides viscous flow, mechanisms such as surface diffusion, Knudsen diffusion, and slip flow can also have a significant effect on gas transport capability. In recent years, the mechanisms and permeability models for the single component gas transport have been widely studied. Beskok et al. have corrected the gas transport equation with considering rarefied gas effect [13]. However, they did not consider the real gas effect. Based on Beskok's model, Michel et al. have introduced real gas effect to the calculation of gas mean free path [14]. Javadpour et al. derived the apparent gas permeability model for single component by through linear superposition of Knudsen diffusion and slip flow [15]. Other mechanisms such as effect of adsorption layer on pore size have not been considered. Guo et al. have conducted the lab experiments for gas flow through nano membranes and proposed corresponding permeability model [16]. A good match between analytical solution and the experimental results was obtained. However, their study is only limited for inorganic pores. Wu et al. constructed the single component gas transport model by re-deriving the weight coefficient of diffusion and slip flow [17]. The surface diffusion and Knudsen diffusion have been considered in their model. Shi et al. have derived the permeability model under different Knudsen number by through nonlinear superposition of diffusion and slip flow. Rahmanian et al. proposed the gas transport model by determining the weight coefficient of diffusion and advection through experiments [18]. Cao et al. have proposed the single component gas permeability model based on capillary model [19]. Mechanisms such as gas adsorption, formation water, slip flow, and free molecular flow have been considered in their model. The main permeability models proposed by previous scholars are summarized as follows:

5.3.1 *Javadpour's Model*

Javadpour et al. established an apparent permeability model considering viscous flow and Knudsen diffusion [15]. The model is as follows:

$$k_a = \left[\frac{2\mu_g M}{3RT\rho_g}\left(\frac{8RT}{\pi M}\right)^{0.5}\frac{8}{r} + \left[1 + \left(\frac{8\pi RT}{M}\right)^{0.5}\frac{\mu_g}{p_g r}\left(\frac{2}{\alpha} - 1\right)\right]\right]k_\infty \tag{5.9}$$

where, ρ_g is the gas density, (kg/m^3); μ_g is the gas viscosity, (Pa s); p_g is the gas pressure, (Pa); k_a represents the apparent permeability with considering Knudsen diffusion and viscous flow, (m^2); k_∞ is the absolute or intrinsic permeability, (m^2); and α is the correction factor; R is the ideal gas constant, its value is 8.314 J/(mol K); M is the gas molar mass, (kg/mol).

5.3.2 *Civan's Model*

Another commonly used model is called the Civan's model, Civan applies the Beskok's model to tight porous medium and considers both viscous flow and Knudsen diffusion. The apparent permeability model proposed by Civan and Vafai [20] is:

$$k_a = k_\infty f(K_n) = k_\infty(1 + \alpha(K_n)K_n)\left(1 + \frac{4K_n}{1 - bK_n}\right) \tag{5.10}$$

where, k_a is the apparent permeability of porous media considering viscous flow and Knudsen diffusion; k_∞ is the intrinsic permeability; K_n is the Knudsen number; α is the dimensionless sparse coefficient; b is the slip coefficient of gas flow in the microtube model, usually taken as − 1.

Based on the Beskok model, Civan gives the dimensionless sparse coefficient α which can be calculated as follows:

$$\alpha = \frac{1.358}{1 + \frac{0.17}{K_n^{0.434}}} \tag{5.11}$$

5.3.3 DGM Model

The dust gas model (DGM) is also used to describe the coupling effect of viscous flow, Knudsen diffusion, and molecular diffusion at the same time. The form of apparent permeability can be expressed as:

$$k_a = k_\infty \left(1 + \frac{b_k}{p_g}\right) \tag{5.12}$$

$$b_k = \frac{D_k^* \mu_g}{k_\infty} \tag{5.13}$$

where, p_g and μ_g respectively represent the pressure and viscosity of the gas; b_k is the Klinkenberg coefficient; and D_k^* is the Knudsen diffusion constant.

5.3.4 Florence's Model

When $Kn \ll 1$, with considering Knudsen diffusion and viscous flow, Florence, Rushing [21] obtained the apparent permeability expression in dense porous media based on the Beskok-Karniadakis model:

$$k_a = k_\infty \left(1 + \frac{4\lambda}{r}\right) \tag{5.14}$$

where, λ is the mean free path of the gas molecule; r is the shale pore radius. The meaning of other parameters can be found from previous equations.

5.3.5 General Slip Regime Model

Zhou et al. [22] have also proposed the permeability model for rarefied gas flow in porous media, which can cover the whole or general slip regimes [22]. Thus, the model is called the general slip regime model or GSR model. The GSR model is derived based on the navier–stokes equation combining 1st order slip boundary condition. The proposed GSR model is as follows:

$$k_a = k_\infty \cdot \left(I + S_1 \sigma \overline{K_n}\right) \cdot \left(I + S_2 \sigma \overline{K_n}\right)^{-1} \tag{5.15}$$

where, S_I and S_2 are geometry dependent rank 2 tensors which can be obtained by fitting the experimental or numerical data to GSR model; I is the unit tensor; σ is the tangential momentum accommodation coefficient; $\overline{K_n}$ is Knudsen number when the pressure approaching zero.

5.3.6 Guo's Model

Based on the Klinkenberg model, our group also obtained the apparent permeability formula for gas transport in nano pores with considering viscous flow, slip flow, and Knudsen diffusion [16]. The expression of the apparent permeability model derived from Guo's model [16] is:

$$k_{app} = k_{\infty}\left(1 + \frac{8\sigma\mu_g}{r\overline{p}}\sqrt{\frac{\pi RT\phi}{M}}\right) \tag{5.16}$$

where, M is the molar mass of the gas, kg/mol; T is the temperature, K; $\overline{P}$ is the average pressure, Pa; μ_g is the viscosity of the gas, Pa s; ϕ is the porosity of the shale, dimensionless; R is the gas constant, 8.314 J/(mol K); r is the pore radius (m).

Also, on the basis of the previous research, we further introduced the improved permeability model with considering three-phase adsorption, pore radius and porosity change due to gas desorption, and the aforementioned mechanisms [23]. The model is shown in Eqs. (17) to (19). For simplification, the meaning of each parameter can be found from Guo et al. (2019).

$$k_{app} = \frac{2\mu r_e M}{3RT\rho}\sqrt{\frac{8RT}{\pi M}} + \left[1 + \sqrt{\frac{8\pi RT}{M}}\frac{\mu}{\overline{P}r_e}\left(\frac{2}{\alpha} - 1\right)\right]\frac{r_e^2}{8} \tag{5.17}$$

$$\phi = \chi \cdot \phi_{\max}\frac{\left\{r_{\max} - d_H \cdot \beta - d_C \cdot \left[\beta\frac{P}{P+P_H} + (1-\beta)\frac{P}{P+P_L}\right]\right\}^2}{r_{\max}^2} \tag{5.18}$$

$$r_e = r_{\max} - d_H \cdot \beta - d_C \cdot \left[\beta\frac{P}{\mathrm{P} + P_H} + (1-\upbeta)\frac{P}{P + P_L}\right] \tag{5.19}$$

However, deficiencies are still existed in the study of gas transport in nano pores. (1) Most research is limited to single component gas transport; (2) The physical meaning of weight coefficient is not clear or there exists empirical coefficient in the weight coefficient which cannot be used for macro-scale flow simulation; (3) Many mechanisms are still not considered, such as real gas effect, water content, adsorption stress, effect of adsorption layer, and difference between organic and inorganic nanopores, etc. For the reliable numerical simulation of TCG transport process during CO_2-ESGR, it is urgently required to propose an improved transport model which can effectively and reasonably combine aforementioned gas transport mechanisms. Therefore, based on our previous research, we tentatively propose the TCG transport model which is shown in Fig. 5.1. As it can be found from Fig. 5.1, firstly, the newly discovered water layer effect should be considered, especially in the shale inorganic pores which are water-wet. Secondly, when the CO_2 is injected into the shale pores, the CO_2 will replace the methane by using competitive adsorption. Thus, the dynamic TCG adsorption model should be considered. Thirdly, during the displacing process,

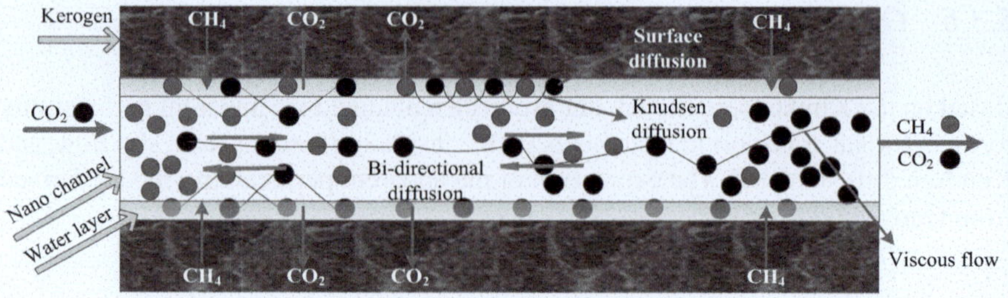

Fig. 5.1 Two component gas transport mechanisms which should be considered in the future study of CO_2-ESGR. The red balls stand for CO_2 molecules and the black balls denote the CH_4 molecules

some mechanisms (not only for pure gas, but also for TCG) should be considered. Generally, these mechanisms include but not limited to: surface diffusion, Knudsen diffusion, viscous flow, molecular diffusion, and the bi-directional diffusion, etc.

5.4 Adsorption Models for Two Component Gases System

5.4.1 Selective Adsorption Mechanism

As stated previously, the adsorption process can be divided into physical adsorption and chemical adsorption. Among them, the adsorption force of physical adsorption is van der Waals force, and the heat of adsorption is small; the adsorption force of chemical adsorption is chemical bond force, and the heat of adsorption is large. In the TCG competitive adsorption process, the gas adsorption is an exothermic reaction. The stronger the adsorption force between gas and adsorbent surface, the larger the adsorption heat, and the more easily the gas is adsorbed. Both the adsorbate (gas) and the adsorbent (shale) affect the adsorption capacity. When the molecules of methane, carbon dioxide, and nitrogen are adsorbed on the same carbon surface, the gas adsorption capacity is different. The shale properties also affect gas adsorption. Ji et al. [24] conducted isothermal adsorption studies for methane on three different types of kerogen. It was found that adsorption heat increased gradually from type I to type III kerogen. The reason may be that the kerogen type promoted the development of aromatic group.

The competition of adsorption sites exists in the process of TCG adsorption, which is called competitive adsorption. The adsorption characteristics of different gases due to their different adsorption capacities are also quite different [25]. The fundamental cause is the selective adsorption capacity of shale for different gases. The selectivity factor or separation coefficient (S) is commonly used to characterize the selective adsorption capacity in TCG adsorption system [26]. The expression of the selectivity factor for the TCG adsorption system is:

$$S = \frac{(x/y)_i}{(x/y)_j} \tag{5.20}$$

where, x and y represent the mole fractions of the adsorbed phase and the free phase for the gas component at the adsorption equilibrium. The selectivity factor S represents the preferential adsorption capacity of the component i in the TCG system [27], and the larger the S, the stronger the preferential adsorption capacity of the component i is.

Zeng et al. have studied the adsorption of pure methane, pure carbon-dioxide, and TCG of methane-carbon dioxide on the organic nanopores in SGRs by using molecular simulation methods [26]. The nanopores are of different geometries and sizes. It was found that under reservoir pressure, the adsorption capacity of carbon dioxide is stronger than those of methane, which is shown in Fig. 5.2 (the selectivity of CO_2 over CH_4). It is apparent that the values of selectivity factor are larger than 1 in all cases, which means that the adsorption capacity of carbon dioxide is greater than methane in the TCG system (CO_2-CH_4) and CO_2 can displace CH_4 effectively.

Chareonsuppanimit, Mohammad [28] studied the adsorption capacity of three gases (CO_2, CH_4 and N_2) in shale, and found that the adsorption capacity of the three gases is CO_2 > CH_4 > N_2. When CH_4 and CO_2 are simultaneously present

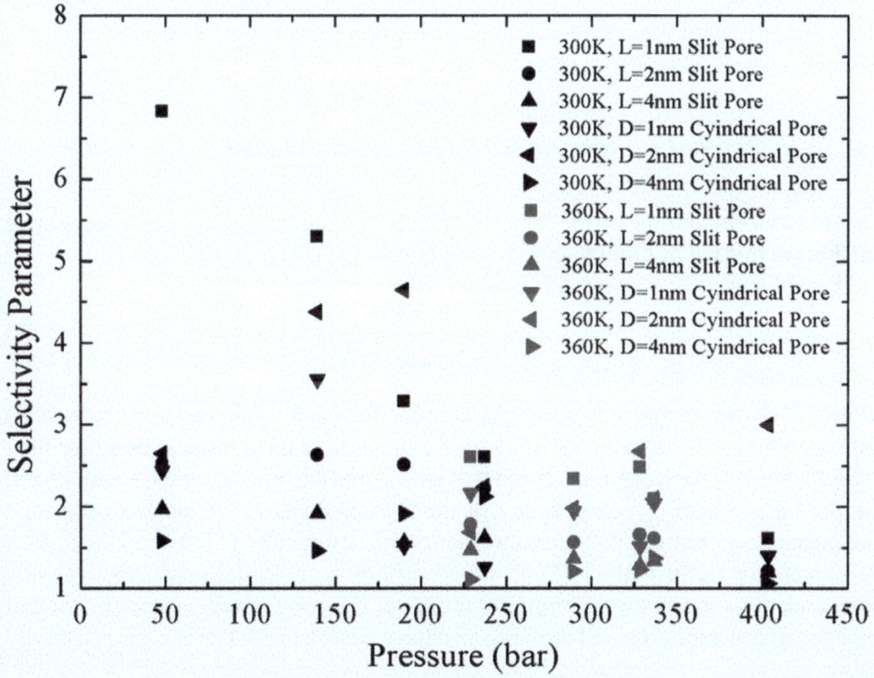

Fig. 5.2 Selectivity of CO_2 over CH_4 at different temperature and pressure in shale nano pores [26]

in the shale, the TCG has a competitive adsorption phenomenon. At the same time, the adsorption characteristics exhibited by the TCG are greatly different from the adsorption of single component gas. Wang [29] conducted single-component adsorption and competitive adsorption studies of CH_4 and C_2H_6 on shale. The experimental results show that in the single component adsorption, the excess adsorption amount increases with pressure within the experimental pressure range. Nevertheless, for the TCG adsorption, the total excess adsorption amount increases firstly, and then decreases at a pressure around 10 MPa. When the adsorption amount reaches the maximum value, this phenomenon is particularly obvious if the concentration of C_2H_6 component is high and the experimental temperature is low. This should be due to the interaction of the two gases in the adsorption phase.

5.4.2 Two-Component Gas Adsorption Model

The competitive adsorption in the TCG system is the fundamental for CO_2-ESGR and CO_2 geo-sequestration. The research group of Nuttall and Kang found that the adsorption capacity of carbon dioxide is 5 to 10 times of methane by experiments [4, 30]. And they believe that the competitive adsorption in CO_2-ESGR is similar as CO_2-ECBM. For the competitive adsorption, many models have been proposed. Here, we presented the commonly used adsorption models for TCG system.

5.4.3 Extended Langmuir Adsorption Model (E-L)

The E-L model is one of the typical models for characterizing gas competitive adsorption for two or more than two components [31]. The expression is:

$$V_i = \frac{(V_L)_i b_i P_i}{1 + \sum_{j=1}^{n_c} b_j P_j} \tag{5.21}$$

where, V_i is the actual adsorbed gas amount for each component; n_c represents the number of gas components; $(V_L)_i$ is the monolayer adsorption volume for the i component; b_i is the Langmuir constant of isothermal adsorption for the i component; the partial pressure of component i in the free phase is P_i; P indicates the total gas pressure; y_i represents the mole fraction of component i. For the E-L model, the selectivity factor is the ratio of the adsorption equilibrium constants, which is independent of gas concentration and pressure. However, studies have shown that changes in concentration and pressure also have a large impact on the gas selectivity factor.

5.4.4 Ideal Adsorption Solution Model

The ideal adsorption solution (IAS) theory assumes that competitive adsorption is similar to ideal solution adsorption which can be characterized by using Raoul's law [32]. The IAS model is as follows:

$$\mathrm{p}y_i = p_i^o(\pi)x_i \tag{5.22}$$

where, p is the total pressure of the mixture; x_i and y_i are the mole fraction of component i in the gas phase and the adsorption phase, respectively; π is the expansion pressure, defined as the amount by which the surface tension of the adsorbate is adsorbed on the surface of the adsorbent; $p_i^o(\pi)$ is the pressure of the component i at the same P–T. The expansion pressure of the one component can be obtained by Gibbs equation. The IAS model depends on the isotherm adsorption equation of single-component gas for calculation. The theoretical prediction results of the ideal adsorption solution also depend on the selection of single-component isotherm adsorption models. According to Clarkson and Bustin [31], it can be found that the prediction accuracy of the same model is quite different if the mixing ratios of the two identical gases are different (Table 5.2).

5.4.5 Lattice Density Functional Model

The lattice density functional model (lattice DFT) [33] not only considers the pore structure of the adsorbent, but also considers the interaction between the adsorbate and the adsorbent. It is assumed that there are N kinds of gases in the slit holes. The N kinds of gas molecules are similar in diameter, and each molecule can exist in the free phase or be adsorbed in the adsorption process. The lattice adsorption sites, or crystal adsorption sites, are marked as j ($j = 1, \ldots, J$) according to their distance from the surface of the adsorbent. The molecular i of a certain component satisfies Eq. (23) when the adsorption/desorption between the free position and the lattice adsorption site j reaches thermodynamic equilibrium:

$$\Delta u_i^j - \mathrm{T}\Delta s_i^j = 0 \tag{5.23}$$

where, Δu_i^j and Δs_i^j represent the energy- and entropy- changes in the exchange process respectively. For N kinds of gases and J lattice adsorption sites, there should be $N \times J$ equations. In Eq. (23), the entropy change can be written as:

$$\Delta s_i^j = kln\frac{\theta_i^j\left(1 - \theta^b\right)}{\theta_i^b\left(1 - \theta^j\right)} \tag{5.24}$$

Table 5.2 Relative error between predicted values and experimental results of each model [31]

Ratio of CH_4 to CO_2	Extended Langmuir			IAS/Langmuir			IAS/D-R			IAS/D-A		
	V_{total} (%)	V_{CH_4} (%)	V_{CO_2} (%)	V_{total} (%)	V_{CH_4} (%)	V_{CO_2} (%)	V_{total} (%)	V_{CH_4} (%)	V_{CO_2} (%)	V_{total} (%)	V_{CH_4} (%)	V_{CO_2} (%)
9:1	9	5	30	8	6	13	8	2	24	7	16	19
3:1	11	28	38	9	14	25	9	25	32	7	7	8

where, k is the Boltzmann constant; θ_i^j is the probability that the molecule i is located on the lattice adsorption site j; θ^j is the probability that any molecule is located on the lattice adsorption site j; θ^b is the probability that any molecule is in the free position; θ_i^b is the probability that the molecule i is in the free position. At the same time, the probability θ can also represent the lattice occupancy ratio, which stands for the ratio of lattice adsorption sites occupied by adsorbate molecules.

5.4.6 *Multi-Region Extended Langmuir Model*

For the multi-region extended Langmuir model (MREL) model, the equation satisfies the conditions of thermodynamic consistency in each zone [34]. When there are n components in the mixture system, the saturated adsorption amount can be calculated by the Langmuir equation if assuming that the system is pure component. We can sort the saturated adsorption amount in order from small to large and then the entire adsorbent region can be divided. For n regions, the model expression is as follows:

$$q_i = \sum_{j=1}^{n} q_{i,j} = \sum_{j=1}^{n} \frac{\left(q_{s,j} - q_{s,j-1}\right) b_i p_i}{1 + \sum_{k=N_j}^{N} b_k p_k} \quad (5.25)$$

where, q_i is the total adsorption amount of component i; $q_{i,j}$ is the amount of adsorption of component i on region j; q_s represents the saturated adsorption amount calculated by the Langmuir equation as a single component; b represents the Langmuir constant; i represents the component; j represents the designated region; $q_{s,j}$ is the saturated adsorption capacity of component i in region j; $q_{s,j-1}$ is the saturated adsorption capacity of component i in region *j-1*; n is the number of regions; N is the number of components; N_j is the number of adsorbed gas components in the region j.

5.4.7 *Laden Ratio Correlation Model (LRC)*

The LRC model is also called the extended Langmuir–Freundlich equation [35]. The LRC model is as follows:

$$\theta_i = \frac{n_i}{n_i^0} = \frac{(b_i p_i)^{q_i}}{1 + \sum_{i=1}^{m} (b_i p_i)^{q_i}} \quad (5.26)$$

where, θ_i is the surface coverage of component i; b_i, q_i, n_i^0 are the parameters obtained by fitting the equilibrium data of pure component i to Langmuir-Freundich equation, specifically b_i is the Langmuir constant of component i; q_i indicates system unevenness parameter; n_i^0 represents the saturated adsorption amount of component i; m is the number of components, i.e. for TCG system, $m = 2$; p_i is the partial pressure for the i component in the gas phase; n_i is the equilibrium adsorption amount.

5.4.8 Dual-Site Langmuir Equation

The Dual-site Langmuir equation can also be extended to describe the equation for the gas adsorption isotherm. Bhadra, Ebner [36] classified the adsorption sites into two categories. The adsorption isotherm equations of the mixed gas can be obtained after the two types of adsorption sites are respectively expanded. However, it is necessary to note that the adsorption site 1 is strong for component A, but not necessarily for component B. Therefore, there are two cases, if site 1 is a strong adsorption site for both A and B, the adsorption amount of each component in the mixture is calculated as follows:

$$q_A = \left(\frac{q^0_{1.A}K_{1.A}Py_A}{1 + K_{1.A}Py_A + K_{1.B}Py_B}\right)_{site1} + \left(\frac{q^0_{2.A}K_{2.A}Py_A}{1 + K_{2.A}Py_A + K_{2.B}Py_B}\right)_{site2} \tag{5.27}$$

$$q_B = \left(\frac{q^0_{1.B}K_{1.B}Py_B}{1 + K_{1.A}Py_A + K_{1.B}Py_B}\right)_{site1} + \left(\frac{q^0_{2.B}K_{2.B}Py_B}{1 + K_{2.A}Py_A + K_{2.B}Py_B}\right)_{site2} \tag{5.28}$$

If site 1 is a strong adsorption site for component A and a weak adsorption site for component B, the expression becomes:

$$q_A = \left(\frac{q^0_{1.A}K_{1.A}Py_A}{1 + K_{1.A}Py_A + K_{2.B}Py_B}\right)_{site1} + \left(\frac{q^0_{2.A}K_{2.A}Py_A}{1 + K_{2.A}Py_A + K_{1.B}Py_B}\right)_{site2} \tag{5.29}$$

$$q_B = \left(\frac{q^0_{1.B}K_{1.B}Py_B}{1 + K_{2.A}Py_A + K_{1.B}Py_B}\right)_{site1} + \left(\frac{q^0_{2.B}K_{2.B}Py_B}{1 + K_{1.A}Py_A + K_{2.B}Py_B}\right)_{site2} \tag{5.30}$$

where, q_i is the total adsorption amount of component i; $q^0_{\mathrm{m}.i}$ is the maximum adsorption amount of component i for adsorption site m; and $K_{\mathrm{m}.i}$ is the equilibrium constant for temperature of component i at adsorption site m; P is the gas preasure; y_A and y_B are the gas-phase mole fractions of components A and B.

5.4.9 Micropore Size Distribution Model

The micropore size distribution (MPSD) model is also called the pore size distribution model, which is proposed by Hu and Do [37]. In MPSD model, the pore size distribution function plus the Lennard–Jones potential energy function is used to represent the energy distribution. By using this method, the heterogeneity of the activated carbon surface can be characterized. The adsorbent is characterized by the pore size distribution function. The MPSD model uses the integral of adsorption isotherm and pore size distribution to describe the adsorption equilibrium.

The expression for the equilibrium adsorption amount of pure component i is:

$$n_i = \int_{r_{\min,i}}^{\infty} n_i(T, P, r) f(r) dr \tag{5.31}$$

where, $r_{\min,i}$ is the smallest pore size at which pure component i can adsorb; n_i is the equilibrium adsorption amount of component i; $f(r)$ is the micropore size distribution function; r represents the pore radius; $n_i(T, P, r)$ represents the local adsorption isotherm, expressed by the Langmuir equation:

$$n_i(r, P, T) = n_i^0(T) \frac{b(r)P}{1 + b(r)P} \tag{5.32}$$

where, $n_i^0(T)$ is the saturated adsorption capacity for the i component; $b(r)$ is the local adsorption potential constant when the pore diameter is r, which is a function of temperature and pore size r:

$$b(r) = b_\infty exp\left(\frac{E(r)}{RT}\right) \tag{5.33}$$

where, b_∞ is the adsorption potential constant when the temperature is infinite; $E(r)$ is the interaction energy between the adsorbed molecule and the pore wall, and its value is the minimum value of the adsorption potential $\varnothing$ of the component i in the pore with the pore diameter r. The calculation of the adsorption potential $\varnothing(r, z)$ is expressed by the Lennard–Jones potential energy function. In the micropores of the parallel slit shape, the expression of the potential energy function depends on the hypothesis of the pore structure, which is often used in the literature.

5.4.9.1 Mixtures Including Supercritical Component Model

The mixtures including supercritical component (MISC) model [38] was constructed with considering different adsorption mechanisms of supercritical components and subcritical components. This adsorption model can predict gases adsorption containing supercritical components. The MISC model is also known as adsorption equilibrium model containing supercritical components. For the pure component adsorption equilibrium of the MISC model, the pore size distribution in the previous MPSD model is used to characterize the heterogeneity of the adsorbent surface potential field. The pore size distribution conforms to the Gamma distribution. The local model describing the adsorption behavior of the class I and II molecules uses the modified Langmuir equation. The adsorption amount of a single component is the sum of the class I and class II molecules, and the equation is:

$$n_i = \int_{r_{\min,i}}^{\infty} (n_{i,I} + n_{i,II}) f(r) dr = \int_{r_{\min,i}}^{\infty} (n^o_{i,I}\theta_{i,I} + n^o_{i,II}\theta_{i,I}\theta_{i,II}) f(r) dr \tag{5.34}$$

where, n_i is the equilibrium adsorption capacity of component i; $n_{i,I}$ and $n_{i,II}$ are the equilibrium adsorption amounts of the class I and II molecules of component i; $r_{\min,i}$ is the minimum pore size at which component i can be adsorbed; and $n^o_{i,I}$, $n^o_{i,II}$ represent the amount of saturated adsorption for class I and II molecules in each component, respectively; $\theta_{i,I}$ and $\theta_{i,II}$ are the surface coverage of class I and II molecules, described by the modified Langmuir equation:

$$\theta_{i,I} = \frac{b_{0,\,i} exp(E(r,i)/RT) exp(-\varsigma) p_i}{1 + b_{0,i} exp(E(r,i)/RT) exp(-\varsigma) p_i} \tag{5.35}$$

where, b_0 represents the pre-exponential factor in the Langmuir constant and ς represents the interaction parameter between the molecules of the adsorbed phase.

5.5 Analysis of Problems in Current Two Component Adsorption Models

For the extended Langmuir equation, if the gas adsorption capacity of each component has a large difference in concentration and pressure change, there will be a significant error in the prediction result of the extended Langmuir equation.

For the ideal adsorption solution (IAS) model, since the IAS theory relies on the isotherm adsorption model of single-component gas, the selection of different one-component isotherm adsorption models will make the IAS theoretical prediction results different. And also, if the mixing ratio of the same gas mixture is different, the prediction accuracy of the IAS model is also different.

For lattice density functional theory model, in the gas adsorption process, not only the adsorption occurs on the surface of kerogen, but also some gases are adsorbed into the organic matter. This leads to the expansion of organic matter and change of its porosity and permeability. However, this effect has not been considered in the lattice density functional theory model. Thus, there will be a certain error between the experimental results and the theoretical values.

For the MREL model, since the saturated adsorption amount of each component in a certain region j is the same value, the Langmuir model in the region satisfies the thermodynamic consistency. However, for the entire adsorption system, the saturation adsorption amount of the mixture is same as the maximum value of the pure component saturated adsorption amount, which is not necessarily reasonable.

For the LRC model, the LRC equation lacks a strict theoretical basis compared to the extended Langmuir equation and can only be used as an empirical correlation. When pressure is low, the pure gas equation of this model does not satisfy Henry's law. However, because the mathematical expression of the model is simple and easy for calculation, the use of the LRC equation is still quite popular in the adsorption study of the gas mixture.

For the Dual-site Langmuir equation, as with the extended Langmuir equation, if the gas adsorption capacity of each component has a large difference with concentration and pressure change, there will be a significant error in the prediction result of the Dual-site Langmuir equation.

For the MPSD model, the pore size distribution is an inherent feature of the adsorbent. As long as the adsorbent is constant, the pore size distribution is fixed and does not change with P and T, regardless of the type of adsorbed molecules. Therefore, in theory, the parameters fitted by the pure components should be the same. However, if the isotherm equation is applied independently to each component, different pore size distribution parameters may be obtained for each component even the adsorbent is the same. This deviates from the fact that the pore size distribution function is an inherent feature of the adsorbent.

Currently, extensive experimental studies have been carried out on pure component adsorption (such as CH_4 or CO_2), and also two component adsorption of CH_4–CO_2 mixture. Many factors affecting gas adsorption have been analyzed. However, the existing adsorption models rarely consider the capillary condensation phenomenon [39] in the nano pores where shale gas is predominately stored. Capillary condensation is a confinement-induced phase transition which has a significant effect on the gas adsorption and recovery in the nanopore. For TCG in confined phase, the phase envelope in the P–T phase diagram is shifted from that of the bulk phase. Also, under confinement state, the pore critical temperature and hysteresis critical temperature depend on the fluid chemistry, pore-wall chemistry, pore size, and pore geometry [40]. How to consider the capillary condensation phenomenon in the TCG system by using a reasonable and efficient method is also a problem which the researchers have to face. Moreover, in most of the simulation work, the TCG adsorption still used the extended Langmuir model, which cannot reflect the real adsorption process. It is suggested to propose or select more reliable two component adsorption model for future CO_2-ESGR simulation. Also, in the process of CO_2-ESGR, pressure change, gas concentration ratio, and water content, etc. have significant effect on competitive adsorption. These factors should be quantitatively analyzed and dynamically embedded in the two component adsorption model so as to realize more accurate simulation and production prediction of CO_2-ESGR.

5.6 Factors Affecting the Adsorption of Two-Component Gas

Many studies have been carried out to investigate the factors affecting the shale adsorption. And it is commonly recognized that shale adsorption capacity is affected by several factors, such as pressure, temperature, thermal maturity, total organic content (TOC), kerogen type, water content, and pore structure characteristics, etc.

The main factors which have a significant effect on shale adsorption capacity are summarized as follows.

5.7 TOC

In shale, the main site of gas adsorption is organic matter. The adsorption gas on organic matter in shale plays a leading role in the total adsorption amount. So, the content of total organic content (TOC) has a great influence on the two-component gas adsorption. According to the study of Myers and Prausnitz [32] for different gas adsorption on shale and coal, the adsorption volume of CO_2, CH_4 and N_2 on coal is 10–30 times more than on shale with the same maturity (Fig. 5.3). Many reasons can lead to this result, such as TOC, inorganic mineral content, pore structure, and etc. According to Busch et al. [41], under unit TOC condition, the adsorption capacity of shale is similar to or slightly lower than that of coal. However, Gasparik et al. [42] found that the adsorption capacity of CH_4 on coal is much higher than that of shale. Prusty [43] also reached similar conclusions. Due to different affinity of organic matter and clay minerals for various gases [24], the selectivity factor between gases has a large difference. Therefore, the level of TOC will not only affect the adsorption of two-component gas in shale, but also affect the selective adsorption of shale to two-component gas.

5.7.1 Pore Structure Characteristics

According to Cui et al. [25], it is known that the molecular size and pore structure characteristics of the adsorbate greatly affect the competitive adsorption of two-component gas. Compared with CH_4 and N_2, CO_2 has a relatively smaller kinetic diameter and can diffuse from macropores to ultra-micropores (< 0.6 nm). However, the kinetic diameters of CH_4 and N_2 are relatively larger and cannot diffuse in the micropores. Yang et al. [44] showed that the selectivity factor of two-component CO_2–CH_4 gas in pores with small pore diameter is larger than that in pores with large pore diameter. And the smaller the pore diameter, the more obvious the selectivity effect of two-component CO_2–CH_4 gas is (Fig. 5.4). It can be easily found from Fig. 5, under the same pressure and mole fraction of CO_2, all the values of selectivity factor when the pore diameter is 14.0 Å is larger the values of 20.6 Å case However, some studies have shown that the selectivity effect of TCG does not always become significant as the pore size becomes smaller and there exists an optimal pore size. Bhatia [45] found that the selectivity behavior of CH_4 and other hydrocarbon gases is the most obvious when the pore diameter is 11.4 Å by studying the binary gas competitive adsorption characteristics on activated carbon for two hydrocarbon systems: CH_4–C_2H_6, and C_3H_8–C_4H_{10}.

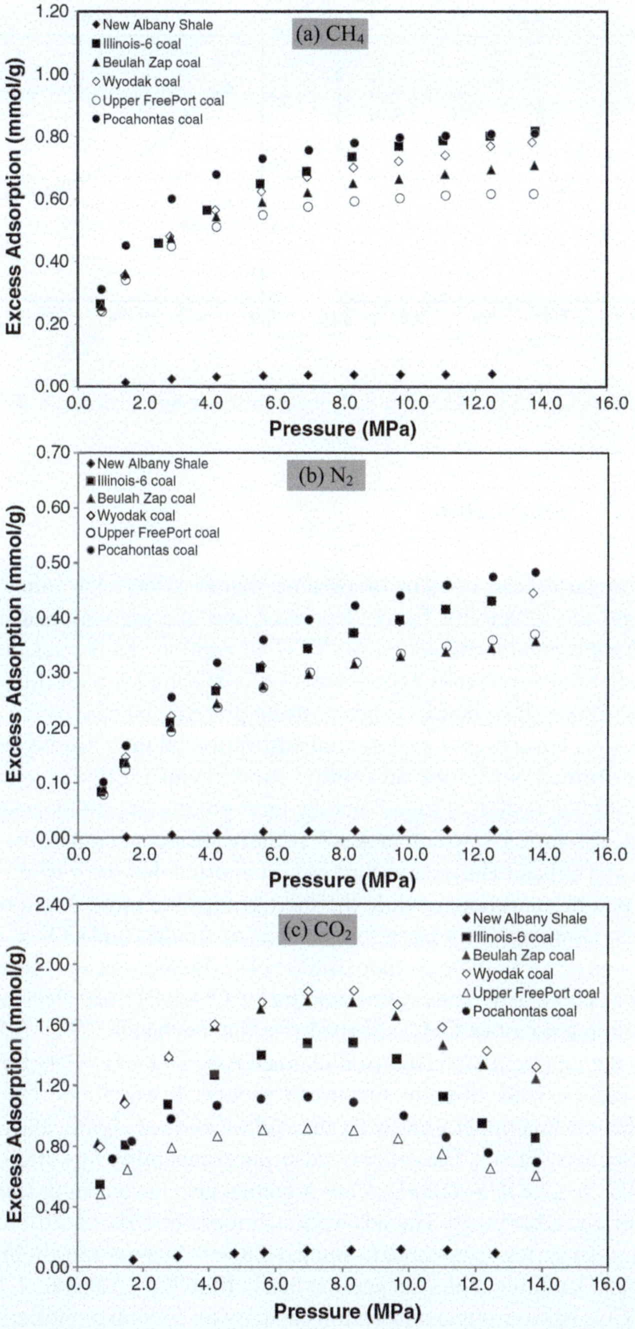

Fig. 5.3 Different gas adsorption on shale and coal. **a** Methane; **b** Nitrogen; **c** Carbon dioxide. After Ref. [28]

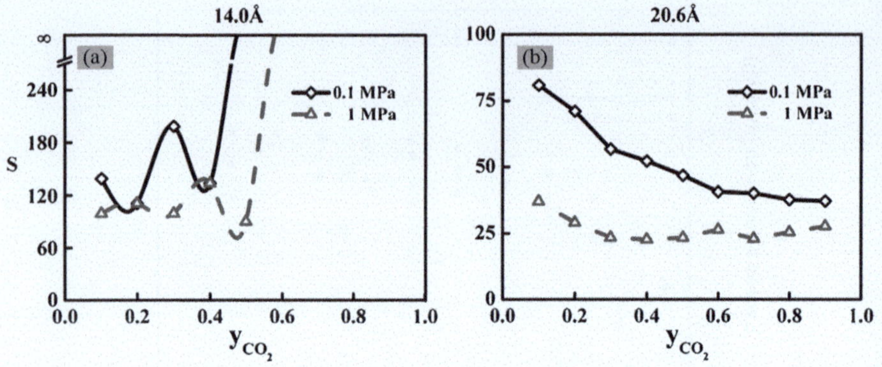

Fig. 5.4 Selectivity factor (S) of CO_2 over CH_4 under different pore diameter. **a** 14.0 Å; **b** 20.6 Å. After Ref. [44]

5.7.2 *Shale Composition*

As stated previously, the content of organic matter affects the amount of TCG adsorption and the selectivity factor. However, the micro-components of organic matter also have a greater impact on the TCG adsorption and the selectivity factor. Previous isothermal adsorption experiments on coal samples with different microscopic compositions show that CO_2 has a strong preferential adsorption on the inert group, while CH_4 has a higher preferential adsorption on the vitrinite group. Busch et al. [41] performed isothermal adsorption experiments on two components CH_4 and CO_2 mixture on coal. It is found that the inert group can provide a large amount of adsorption sites for CO_2, which makes CO_2 have stronger competitive adsorption. Unsworth et al. [46] and Harris and Yust [47] suggested that the vitrinite groups are more developed in micropores, while the inert groups are more developed in intermediate pores. However, the adsorption mechanism of CH_4 and CO_2 in the vitrinite and inert groups remains to be further studied [48]. Weniger et al. [49] showed that as the vitrinite content increases, the amount of CO_2 adsorbed decreases significantly, while the amount of CH_4 adsorbed remains unchanged (Fig. 5.5). Lu et al. [50] studied the competitive adsorption characteristics of CO_2-CH_4 binary gas on nanoporous carbon with different functional groups. It was found that the difference of molecular functional groups on the carbon surface significantly affects the CO_2-CH_4 selectivity factor. The order of adsorption capability for different function group is: $-NH_2$ >–COOH>–OH>H. Clay minerals also have a significant effect on the amount of gas adsorption. The adsorption amount of CH_4 on different types of clay minerals also varies, from large to small, followed by montmorillonite, smectite mixed minerals, kaolinite, chlorite, and the last is Illite [24]. Yang et al. [44] further showed that CO_2 has a stronger adsorption capacity on sodium montmorillonite than CH_4. Duan et al. [51] believe that clay minerals have stronger selective adsorption to CO_2 than other components in shale. And the increase of clay mineral content will increase the selectivity factor of CO_2–CH_4 binary gas.

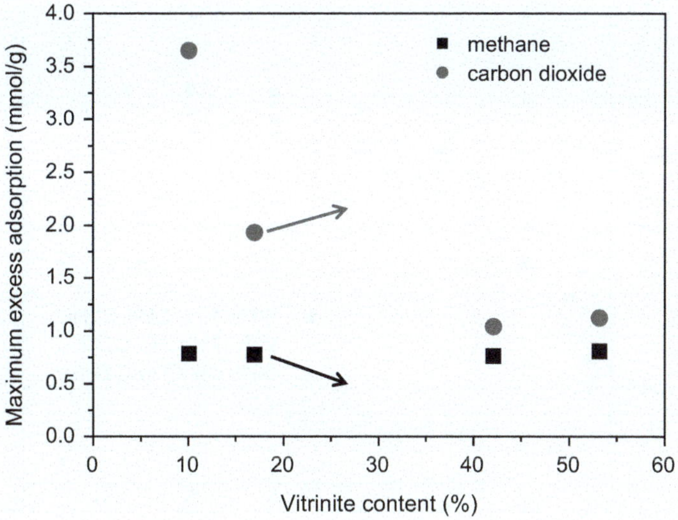

Fig. 5.5 Effect of vitrinite content on the maximum excess adsorption capacity of CO_2 and CH_4 [49]

5.7.3 *Reservoir Pressure*

Busch, Gensterblum [41] believed that pressure was the main driving force for preferential CO_2 adsorption in CO_2-CH_4 mixture. Hartman et al. [52] studied the selectivity factor of binary gas (CH_4 and C_2H_6) under different pressure conditions and found that when the ratio of CH_4 to C_2H_6 is 1:1, the selectivity factor of C_2H_6 under 27.58 MPa is more than 35% higher than that under 6.89 MPa. Shimada et al. [53] found that the adsorption capacity ratio for carbon dioxide, methane, and nitrogen under 1.2 MPa was 8.5: 3.5: 1 at 1.2 MPa. And the ratio under 6 MPa is 5.5: 2: 1. Wang et al. [29] showed that for the adsorption of binary gas (CH_4 and C_2H_6) on shale, the total adsorption capacity increased with the increase of pressure, reached to the maximum value at 10 MPa, and then decreased with the increase of pressure (Fig. 5.6).

5.7.4 *Maturity*

The main adsorption site of two component gases (TCG) in shale is organic matter. When the maturity is continuously increased, the aromatic structure of kerogen is continuously condensed, and the degree of aromatization is continuously enhanced. The oxygen-containing functional group is broken and the affinity for the TCG changes. Also, as maturity increases, the shale pore structure undergoes constant changes. At low maturity, compaction plays a dominant role. At this stage, organic pores are not developed, and mesopores (2–50 nm) and macropores (> 50 nm) in

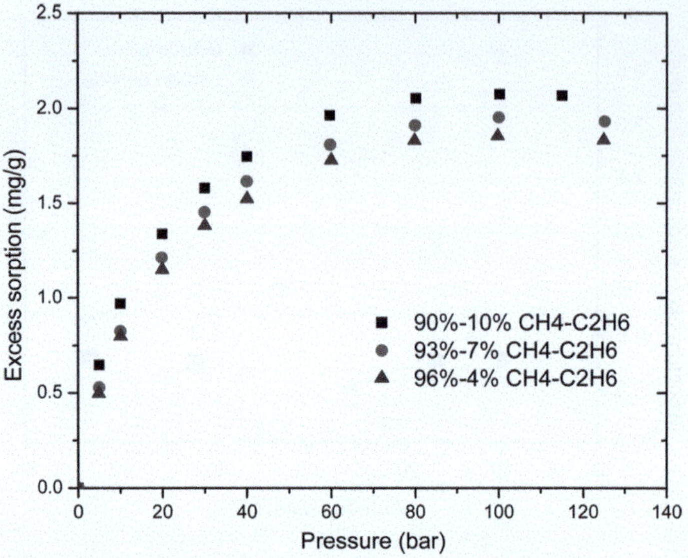

Fig. 5.6 Sorption isotherms of CH_4–C_2H_6 mixtures on shale at 60 °C [29]

shale are decreasing. When the maturity is near the oil window, the original pores are blocked due to the large amount of liquid hydrocarbons generated. At this stage, micropores (< 2 nm), mesopores and macropores are continuously decreasing. When maturity is close to the liquid hydrocarbon cracking stage, the blocked pores are released, and the residual asphalt and kerogen also begin to crack. At this stage, the micropores and mesopores in the shale are increasing. The pore volume and specific surface area are continuously increasing. Therefore, at this stage, the amount of gas adsorption has a tendency to increase with maturity, which is similar to that observed in coal. At the maturity stage (Ro > 3.5%), the organic pores will be destroyed and the shale pore volume and specific surface area will be reduced. Changes in the fraction of organic pores and clay minerals will inevitably change the adsorption and selectivity factor of TCG in shale. Results from Weniger et al. [49] showed that in the range of Ro < 3.0%, the CO_2 adsorption decreases significantly with the increase of maturity, but the decrease of CH_4 adsorption is very tiny (Fig. 5.7).

5.7.5 *Moisture Content*

The effect of moisture content on the competitive adsorption of TCG in shale is also significant. When the shale contains liquid water, the gas molecules not only move at the gas–liquid interface, but also move at the solid–gas interface. Due to the presence of liquid water, the pores are blocked, directly affecting the amount of gas adsorbed. Weniger et al. [49] found that there are a large number of clay minerals in

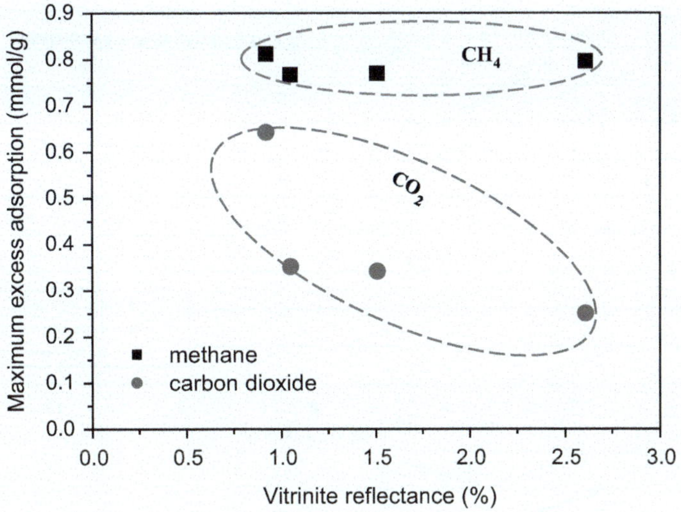

Fig. 5.7 Effect of maturity on the maximum excess adsorption capacity of CO_2 and CH_4 [49]

the shale, and the clay minerals have strong hydrophilic ability, which is obviously different from coal. Therefore, the presence of liquid water has a notable effect on the selective adsorption of TCG. In addition, the presence of water also reduces the preferential adsorption capacity [31] of CO_2, a phenomenon that is particularly pronounced in the inert group. Busch et al. [41] found that when water is contained in shale, water will occupy the adsorption site of CO_2, which makes CO_2 difficult to adsorb (Fig. 5.8). As shown in Fig. 5.8, the y axis means the mole fraction of CO_2 in the free gas (non-adsorbed) phase. The dotted lines denote the mole fraction of the source gas. If the mole fraction of CO_2 is lower than the dotted line (Fig. 5.8a for the dry shale), it means that CO_2 in the free gas phase decreases, i.e. more CO_2 is adsorbed. Otherwise, if the mole fraction of CO_2 is above than the dotted line (Fig. 5.8b for the wet shale), it means that CO_2 as the free gas phase increases, i.e. less CO_2 is adsorbed. It can be concluded that when the shale is wet, CO_2 is becoming difficult to adsorb on the shale. At the same time, CO_2 has a higher solubility in water compared with gases such as CH_4 and N_2. When liquid water exists on the surface of the shale, CO_2 not only participates in competitive adsorption, but also partially dissolves in liquid water, making its competitive adsorption relationship with other gases more complicated.

5.7.6 *Gas Concentration Ratio*

For the competitive adsorption of TCG, the gas concentration ratio also has effect on the gas adsorption. Ma et al. carried out isothermal adsorption experiments with

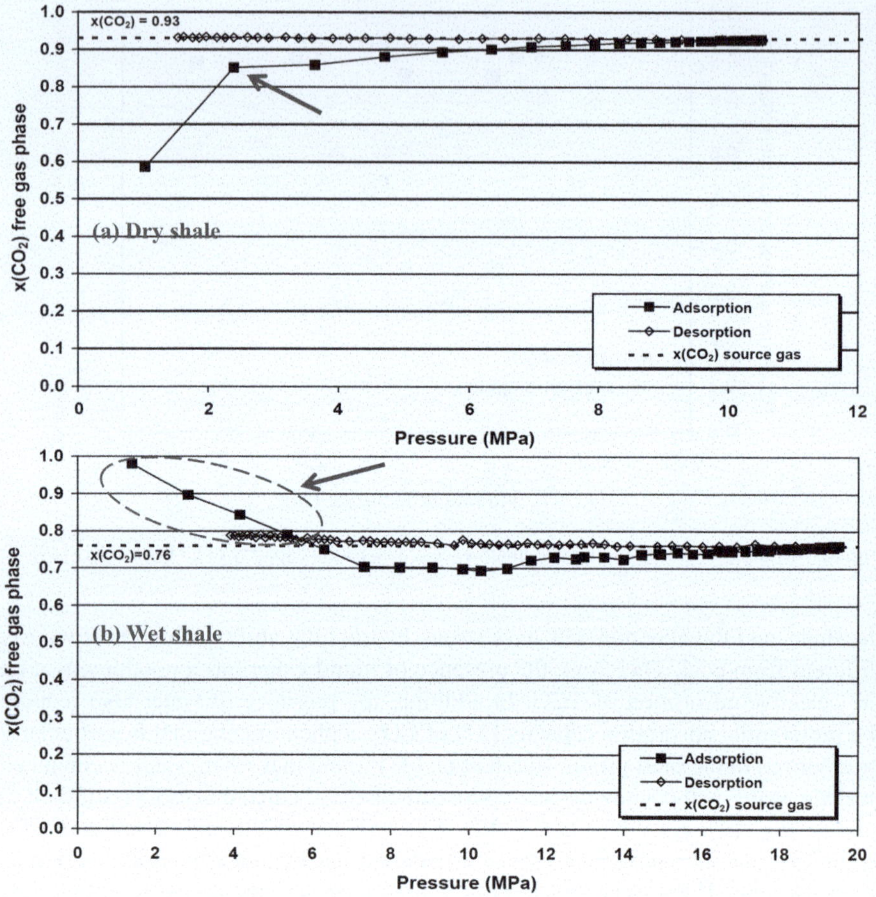

Fig. 5.8 Adsorption measurements on shale samples with different water content [41]. **a** dry shale; **b** moist shale

different CH_4 and CO_2 concentration ratios. It can be found from Fig. 5.9, when comparing Fig. 5.9a, c for the same sample 1, the amount of adsorption decreases as the ratio of CH_4 to CO_2 increases under the same temperature and pressure condition. The same conclusion can be obtained by comparing Figs. 5.9b, d for the same sample 2. In the CO_2-ESGR process, with the injection of carbon dioxide from the injection well and the production of methane and carbon dioxide from the production well, the gas concentration ratio of CO_2 to CH4 is changing with production time. Thus, this dynamic adsorption effect should also be embedded into current numerical simulation models of CO_2-ESGR.

According to the review, it can be found that above mentioned factors have already been studied in previous research by using experimental methods. However, how to effectively combine such factors in the dynamic production process of CO_2-ESGR

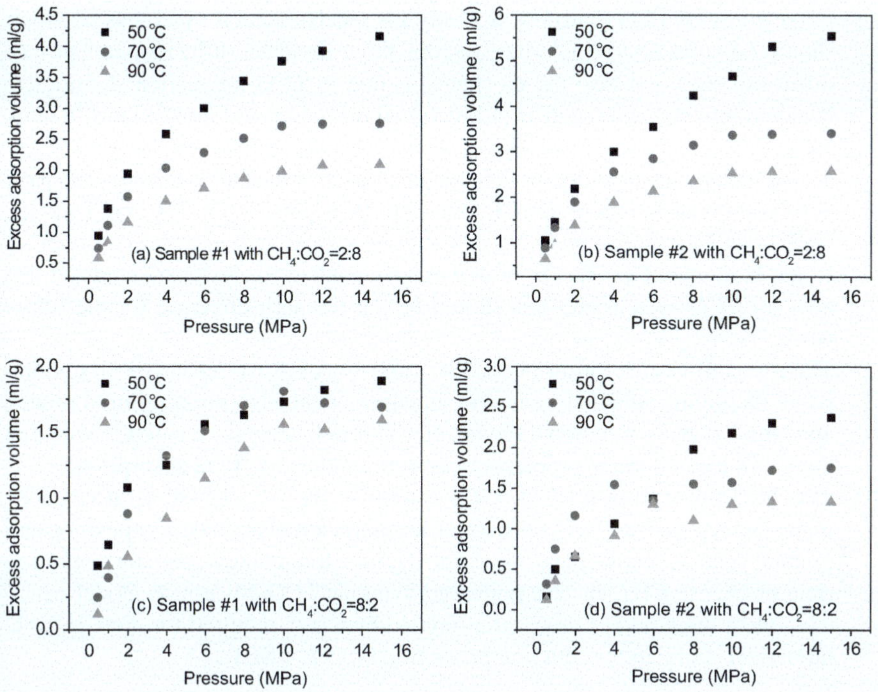

Fig. 5.9 The isothermal adsorption curve of mixed gases on shales under different gas concentration ratio [54]. **a** Sample 1 with CH_4:CO_2 = 2:8; **b** Sample 2 with CH_4:CO_2 = 2:8; **c** Sample 1 with CH_4:CO_2 = 8:2; **d** Sample 2 with CH_4:CO_2 = 8:2

still remains to be a problem and has rarely been considered in the numerical simulation of CO_2-ESGR. During the CO_2-ESGR process, the pressure, gas concentration, pore radius, water content, etc. will change with time. For more accurate numerical simulation of CO_2-ESGR, it is necessary to conduct further research on the development of dynamic TCG adsorption model.

5.8 Remark

In summary, two component gases (TCG) transport in shale nano pores are different from the single component gas transport. Although many efforts have been made for pure gas transport in shale nano pores and also some work have been conducted in the study of CO_2-ESGR by using commercial software, rare work have been conducted for TCG transport in shale gas nano pores and many mechanisms cannot be accurately reflected. Therefore, the key issues which should be considered in the study of CO_2-ESGR have been reviewed. The main findings from this review are as follows:

(1) In the study of TCG transport, not only the mechanisms for pure gas transport should be considered (including viscous flow, Knudsen diffusion, and surface diffusion, etc.), but also the mechanisms due to TCG interaction should also be taken into account (such as bidirectional diffusion, and molecular diffusion, etc.).

(2) The apparent permeability models for gas transport in shale gas reservoirs have been comprehensively reviewed, including Javadpour's model, Civan's model, GSR model, etc. It can be found that although apparent permeability model for single component gas has been extensively studied, there is rare study in this area for TCG system and this should be the research direction for the future work.

(3) The competitive adsorption mechanisms and models for TCG have been summarized, including extended Langmuir equation, ideal adsorption solution model, and lattice density functional model, etc. Through analysis, it can be found that there still existed many problems in current models for TCG adsorption.

(4) The TCG adsorption for CO_2-CH_4 system in the CO_2-ESGR is affected by many factors, such as TOC, pore structure characteristics, shale composition and mineral type, reservoir pressure, maturity, moisture content, and concentration ratio of CO_2 to CH_4, etc. An improved dynamic TCG adsorption model which can reflect the effect of time-variation parameters is required for more reliable numerical study and production prediction of CO_2-ESGR.

References

1. Wei, X., et al. (2007). Numerical simulation of multicomponent gas diffusion and flow in coals for CO_2 enhanced coalbed methane recovery. *Chemical Engineering Science, 62*(16), 4193–4203.
2. Fathi, E., & Akkutlu, I. Y. (2008). Counter-diffusion and competitive adsorption effects during CO_2 injection and coalbed methane production. In *SPE Annual Technical Conference and Exhibition*? SPE.
3. Sun, X., et al. (2016). Numerical simulation of gas injection enhanced coalbed methane recovery based on bi-dispersing diffusion model with Maxwell-Stefan equation. *Journal of China University Petrol, 40*(3), 113–120.
4. Kang, S. M., et al. (2011). Carbon dioxide storage capacity of organic-rich shales. *SPE Journal, 16*(04), 842–855.
5. Brunauer, S., et al. (1940). On a theory of the van der Waals adsorption of gases. *Journal of the American Chemical society, 62*(7), 1723–1732.
6. Choi, J.-G., Do, D., & Do, H. (2001). Surface diffusion of adsorbed molecules in porous media: Monolayer, multilayer, and capillary condensation regimes. *Industrial & Engineering Chemistry Research, 40*(19), 4005–4031.
7. Javadpour, F., Fisher, D., Unsworth, M. (2007). Nanoscale gas flow in shale gas sediments. *Journal of Canadian Petroleum Technology, 46*(10).
8. Burnett, D. (1935). P. rot
9. Hornyak, G. L., et al. (2008). *Introduction to nanoscience and nanotechnology*. CRC Press.
10. Bhattacharya, D., & Lie, G. (1991). Nonequilibrium gas flow in the transition regime: A molecular-dynamics study. *Physical Review A, 43*(2), 761.

11. Karniadakis, G. E., Beskok, A., & Gad-el-Hak, M. (2002). Micro flows: Fundamentals and simulation. *Applied Mechanics Reviews, 55*(4), B76–B76.
12. Hwang, S. T., & Kammermeyer, K. (1966). Surface diffusion in microporous media. *The Canadian Journal of Chemical Engineering, 44*(2), 82–89.
13. Beskok, A., & Karniadakis, G. E. (1999). Report: A model for flows in channels, pipes, and ducts at micro and nano scales. *Microscale Thermophysical Engineering, 3*(1), 43–77.
14. Michel, G., et al. (2011) Parametric investigation of shale gas production considering nano-scale pore size distribution, formation factor, and non-Darcy flow mechanisms. In *SPE Annual Technical Conference and Exhibition*? SPE.
15. Javadpour, F. (2009). Nanopores and apparent permeability of gas flow in mudrocks (shales and siltstone). *Journal of Canadian Petroleum Technology, 48*(08), 16–21.
16. Guo, C., et al. (2015). Study on gas flow through nano pores of shale gas reservoirs. *Fuel, 143*, 107–117.
17. Wu, K., et al. (2016). A unified model for gas transfer in nanopores of shale-gas reservoirs: Coupling pore diffusion and surface diffusion. *SPE Journal, 21*(05), 1583–1611.
18. Rahmanian, M. R., Aguilera, R., & Kantzas, A. (2011). A new unified diffusion-viscous flow model based on pore level studies of tight gas formations. In *SPE Canada Unconventional Resources Conference*? SPE.
19. Cao, C., Li, T., & Fan, B. (2015). Composite permeability model of shale matrix with the effect of adsorption. In *2015 International Conference on Test, Measurement and Computational Methods*. Atlantis Press.
20. Civan, F., & Vafai, K. (2010). A review of approaches for describing gas transfer through extremely tight porous media. In *AIP Conference Proceedings*.
21. Florence, F. A., et al. (2007). Improved permeability prediction relations for low-permeability sands. In *SPE Rocky Mountain Petroleum Technology Conference/Low-Permeability Reservoirs Symposium*. SPE.
22. Zhou, B., et al. (2016). General slip regime permeability model for gas flow through porous media. *Physics of Fluids, 28*(7).
23. Guo, C., et al. (2019). An improved transport model of shale gas considering three-phase adsorption mechanism in nanopores. *Journal of Petroleum Science and Engineering, 182*, 106291.
24. Ji, L., et al. (2012). Experimental investigation of main controls to methane adsorption in clay-rich rocks. *Applied Geochemistry, 27*(12), 2533–2545.
25. Cui, X., Bustin, R. M., & Dipple, G. (2004). Selective transport of CO_2, CH_4, and N_2 in coals: Insights from modeling of experimental gas adsorption data. *Fuel, 83*(3), 293–303.
26. Zeng, K., et al. (2018). Molecular simulation of carbon dioxide and methane adsorption in shale organic nanopores. *Energy & Fuels, 33*(3), 1785–1796.
27. Ruthven, D. M. (1984). *Principles of adsorption and adsorption processes*. Wiley.
28. Chareonsuppanimit, P., et al. (2012). High-pressure adsorption of gases on shales: Measurements and modeling. *International Journal of Coal Geology, 95*, 34–46.
29. Wang, Y., Tsotsis, T. T., & Jessen, K. (2015). Competitive sorption of methane/ethane mixtures on shale: Measurements and modeling. *Industrial & Engineering Chemistry Research, 54*(48), 12187–12195.
30. Nuttal, B. C., et al. (2005). Analysis of Devonian black shales in Kentucky for potential carbon dioxide sequestration and enhanced natural gas production. In *Greenhouse Gas Control Technologies* (Vol. 7, pp. 2225–2228). Elsevier.
31. Clarkson, C., & Bustin, R. (2000). Binary gas adsorption/desorption isotherms: Effect of moisture and coal composition upon carbon dioxide selectivity over methane. *International Journal of Coal Geology, 42*(4), 241–271.
32. Myers, A. L., & Prausnitz, J. M. (1965). Thermodynamics of mixed-gas adsorption. *AIChE Journal, 11*(1), 121–127.
33. Ottiger, S., et al. (2008). Measuring and modeling the competitive adsorption of CO_2, CH_4, and N_2 on a dry coal. *Langmuir, 24*(17), 9531–9540.

34. Bai, R., & Yang, R. T. (2001). A thermodynamically consistent Langmuir model for mixed gas adsorption. *Journal of Colloid and Interface Science, 239*(2), 296–302.
35. Malek, A., & Farooq, S. (1996). Comparison of isotherm models for hydrocarbon adsorption on activated carbon. *AIChE Journal, 42*(11), 3191–3201.
36. Bhadra, S. J., Ebner, A. D., & Ritter, J. A. (2012). On the use of the dual process Langmuir model for predicting unary and binary isosteric heats of adsorption. *Langmuir, 28*(17), 6935–6941.
37. Hu, X., & Do, D. D. (1996). Effect of pore size distribution on the prediction of multicomponent adsorption equilibria. In *Fundamentals of Adsorption: Proceedings of the Fifth International Conference on Fundamentals of Adsorption*. Springer.
38. Zhou, L., et al. (2005). Prediction of multicomponent adsorption equilibrium of gas mixtures including supercritical components. *Chemical engineering science, 60*(11), 2833–2844.
39. Barsotti, E., et al. (2016). A review on capillary condensation in nanoporous media: Implications for hydrocarbon recovery from tight reservoirs. *Fuel, 184*, 344–361.
40. Horikawa, T., Do, D., & Nicholson, D. (2011). Capillary condensation of adsorbates in porous materials. *Advances in colloid and interface science, 169*(1), 40–58.
41. Busch, A., et al. (2006). Investigation of high-pressure selective adsorption/desorption behaviour of CO_2 and CH_4 on coals: An experimental study. *International Journal of Coal Geology, 66*(1–2), 53–68.
42. Gasparik, M., et al. (2014). First international inter-laboratory comparison of high-pressure CH_4, CO_2 and C_2H_6 sorption isotherms on carbonaceous shales. *International Journal of Coal Geology, 132*, 131–146.
43. Prusty, B. K. (2008). Sorption of methane and CO_2 for enhanced coalbed methane recovery and carbon dioxide sequestration. *Journal of Natural Gas Chemistry, 17*(1), 29–38.
44. Yang, N., Liu, S., & Yang, X. (2015). Molecular simulation of preferential adsorption of CO_2 over CH_4 in Na-montmorillonite clay material. *Applied Surface Science, 356*, 1262–1271.
45. Bhatia, S. K. (1998). Adsorption of binary hydrocarbon mixtures in carbon slit pores: A density functional theory study. *Langmuir, 14*(21), 6231–6240.
46. Unsworth, J. F., Fowler, C. S., & Jones, L. F. (1989). Moisture in coal: 2. Maceral effects on pore structure. *Fuel, 68*(1), 18–26.
47. Harris, L. A., & Yust, C. S. (1976). Transmission electron microscope observations of porosity in coal. *Fuel, 55*(3), 233–236.
48. Lamberson, M. N., & Bustin, R. M. (1993). Coalbed methane characteristics of gates formation coals, northeastern British Columbia: Effect of maceral composition. *AAPG Bulletin, 77*(12), 2062–2076.
49. Weniger, P., et al. (2010). High-pressure methane and carbon dioxide sorption on coal and shale samples from the Paraná Basin, Brazil. *International Journal of Coal Geology, 84*(3–4), 190–205.
50. Lu, X., et al. (2015). Competitive adsorption of a binary CO_2–CH_4 mixture in nanoporous carbons: Effects of edge-functionalization. *Nanoscale, 7*(3), 1002–1012.
51. Duan, S., et al. (2016). Adsorption equilibrium of CO_2 and CH_4 and their mixture on Sichuan Basin shale. *Energy & Fuels, 30*(3), 2248–2256.
52. Hartman, R. C., et al. (2011). Shale gas-in-place calculations part II—Multi-component gas adsorption effects. In *SPE Unconventional Resources Conference/Gas Technology Symposium*. SPE.
53. Shimada, S., et al. (2005). Displacement behavior of CH_4 adsorbed on coals by injecting pure CO_2, N_2, and CO_2–N_2 mixture. *Environmental Geology, 49*, 44–52.
54. Ma, Y., et al. (2019). Study of CH_4 and CO_2 competitive adsorption on shale in Yibin, Sichuan Province of China. *Carbon Resources Conversion, 2*(1), 35–42.

Chapter 6
Two Component Gas Transport Model in Shale Gas Nanopores

Due to the shortage of conventional oil and gas resources, unconventional shale gas resources are becoming increasingly important so as to ensure the supply of global oil and gas resources. Shale gas mainly exists in organic shale and its interlayer, which are with ultra-low permeability and porosity. At present, carbon dioxide displacement is the most commonly used method to exploit shale gas. This method can not only improve the recovery of shale gas by replacing methane by adsorption, but also store a large amount of carbon dioxide in shale reservoir, thus slowing down greenhouse effect. To study the shale gas displacement by carbon dioxide, it is necessary to study the gas flow mechanism. In this chapter, on the basis of predecessors, considering various possible flow mechanisms, an apparent permeability model is established to describe the flow characteristics of methane-carbon dioxide two-component gas. Finally, the influence factors on two component gas permeability are analyzed to reveal the characteristics of apparent permeability.

6.1 Two-Component Gas Flow Mechanisms and Model

In the aspect of mechanism research, Lee and Sidle [1] studied the desorption of methane during the process of shale gas production by using molecular dynamics (MD) simulation. Zhai et al. [2] studied the adsorption properties of methane in shale reservoir by the same method. Zhang and Cao [3] studied the replacement efficiency of carbon dioxide on methane under different formation depth. Sui and Yao [4] studied the effects of different chemical components on the competitive adsorption between carbon dioxide and methane. Wu et al. [5] used MD method to simulate the process of driving methane in carbon nanotubes and reached the conclusion that carbon dioxide can directly drive adsorbed methane. Yuan et al. [6] used MD method to simulate the process of injecting carbon dioxide into the surface of graphene to

C. Guo and Z. Yang, *Shale Gas Production: Concept, Models, and Techniques*, Petroleum Engineering, https://doi.org/10.1007/978-3-031-84552-9_6

drive methane, and found that carbon dioxide is easier to adhere to the surface of graphene.

In the area of improving shale gas production by using carbon dioxide injection, Busch et al. [7] studied the carbon dioxide storage potential of Muderong shale samples and different clay minerals by diffusion and gas adsorption experiments. Ye et al. [8] used COMSOL to simulate the replacement of methane by carbon dioxide in shale. They classified the process into three stages: gas depressurization stage, carbon dioxide adsorption stage, and carbon dioxide and methane co-production stage. Hoffman and Shoaib [9] used numerical simulations to evaluate different EOR methods for reservoirs, including carbon dioxide flooding and cyclic injection, to develop Middle Bakken areas. Sanchez-Rivera et al. [10] studied the main parameters affecting the carbon dioxide huff and puff process of Bakken formation through numerical simulation.

6.2 Model Assumptions

(1) There are only two-component gases in the shale pores, that is, methane and carbon dioxide.
(2) There are both adsorption and free gas in shale nanopores.
(3) The ratio of two-component gas content is the same everywhere in shale pores.
(4) Ignore the influence of other impurities in shale pores.

6.3 Gas Mass Flow Equation

Adsorption gas on the surface of organic nano-pores of shale is the main gas fraction in shale nanopores. The competitive adsorption between two-component gases occur in the organic matter nanopores, which can be obtained by extending the Langmuir isothermal adsorption model [11]:

$$V_1 = \frac{V_{L1} \cdot P_1}{P + P_{L1}},$$
$$V_2 = \frac{V_{L2} \cdot P_2}{P + P_{L2}} \tag{6.1}$$

where, V_1 and V_2 are the adsorption volume of methane and carbon dioxide, m^3/kg; V_{L1} and V_{L2} are the Langmuir maximum adsorption volume of methane and carbon dioxide, m^3/kg; P_{L1} and P_{L2} are the Langmuir pressure constant of methane and carbon dioxide, Pa; P_1 and P_2 are the partial pressure of methane and carbon dioxide, Pa; P is the total gas pressure, Pa. For each component gas, the surface coverage can be obtained as follows:

$$\theta_1 = \frac{P_1}{P + P_{L1}},$$
$$\theta_2 = \frac{P_2}{P + P_{L2}}, \tag{6.2}$$

where, θ_1 and θ_2 are the surface coverage of methane and carbon dioxide, dimensionless. Shale organic nanopores have a strong adsorption to gas molecules, therefore there will be a large number of methane and carbon dioxide gas molecules adsorbed on the surface of organic nanopores. This will leads to the change in pore radius. Thus, the pore radius needs to be modified:

$$r_t = r_m - d_1 \frac{P_1}{P + P_{L1}} - d_2 \frac{P_2}{P + P_{L2}} \tag{6.3}$$

where, r_t is the corrected pore radius, m; r_m is the pore radius before correction, m; d_1 is the molecular diameter of methane, m; d_2 is the molecular diameter of carbon dioxide gas. The gas density can be calculated by using following equations:

$$\rho_1 = \frac{P_1 M_1}{ZRT},$$
$$\rho_2 = \frac{P_2 M_2}{ZRT} \tag{6.4}$$

where, M_1 and M_2 are the molar mass of methane and carbon dioxide, kg/mol; Z is the gas compression factor, dimensionless; R is general gas constant, 8.314 J/ (mol K); T is temperature, K. And the gas viscosity also needs to be modified since it is affected by pressure and other factors:

$$\mu_1 = \mu_1^0 \frac{1}{1 + \lambda \cdot K_{n1}},$$
$$\mu_2 = \mu_2^0 \frac{1}{1 + \lambda \cdot K_{n2}} \tag{6.5}$$

where, μ_1^0 and μ_2^0 are the initial gas viscosity of methane and carbon dioxide, Pa s; λ is the gas viscosity parameter, dimensionless; K_{n1} and K_{n2} are the Knudsen number of methane and carbon dioxide, dimensionless. The calculation equation of gas Knudsen number is listed as follows:

$$K_{n1} = \frac{k_b T}{2\sqrt{2}\pi d_1^2 P_1 r_t},$$
$$K_{n2} = \frac{k_b T}{2\sqrt{2}\pi d_2^2 P_2 r_t} \tag{6.6}$$

where, k_b is Boltzmann constant, 1.38×10^{-23} J/K. The partial pressure of each component (P_1 and P_2) is calculated as follows:

$$P_1 = P \cdot x,$$
$$P_2 = P \cdot y = P \cdot (1 - x) \tag{6.7}$$

where, x and y are the volume fraction of methane and carbon dioxide, dimensionless. At the same time, in the actual production, the gas component ratio can also be used to express the component fraction. Here, we defined an important parameter: methane-carbon dioxide component ratio (MCCR), which is as follows:

$$\varepsilon_{ij} = \frac{x}{y} = \frac{x}{1-x},$$
$$\varepsilon_{ji} = \frac{y}{x} = \frac{y}{1-y} \tag{6.8}$$

where, ε_{ij} is methane-carbon dioxide component ratio (MCCR), dimensionless; ε_{ji} is carbon dioxide-methane component ratio, dimensionless. The transformation relationship between the two is as follows:

$$\varepsilon_{ij} \cdot \varepsilon_{ji} = 1 \tag{6.9}$$

The mass flow equation caused by the slippage modified viscous flow is as follows:

$$J_{v1} = -\frac{\phi \rho_1 r_t^2}{8\mu_1}(1 + \alpha \cdot K_{n1})\left(1+\frac{4 \cdot K_{n1}}{1 - b \cdot K_{n1}}\right)\nabla P_1,$$
$$J_{v2} = -\frac{\phi \rho_2 r_t^2}{8\mu_2}(1 + \alpha \cdot K_{n2})\left(1+\frac{4 \cdot K_{n2}}{1 - b \cdot K_{n2}}\right)\nabla P_2 \tag{6.10}$$

where, ϕ is porosity, dimensionless; b is the fitting constant of slippage effect, dimensionless; α is the fitting function of slippage effect, dimensionless:

$$\alpha = \alpha_0 \tan^{-1}\left(4K_n^{0.4}\right) \tag{6.11}$$

where, α_0 is the equation fitting constant, dimensionless. The mass flow equation caused by Knudsen diffusion can be obtained as follows:

$$J_{k1} = -\frac{M_1 D_{k1}}{RT}\nabla P_1,$$
$$J_{k1} = -\frac{M_2 D_{k2}}{RT}\nabla P_2 \tag{6.12}$$

where, D_{k1} and D_{k2} are the Knudsen diffusion coefficient of methane and carbon dioxide, m^2/s. The Knudsen diffusion coefficient can be obtained as follows:

$$D_{k1} = \frac{2r_t}{3}\sqrt{\frac{8RT}{\pi M_1}},$$

$$D_{k2} = \frac{2r_t}{3}\sqrt{\frac{8RT}{\pi M_2}} \tag{6.13}$$

The mass flow flux caused by molecular diffusion is as follows:

$$\begin{aligned} J_{m1} &= -\frac{M_1 D_{m12}}{RT}\nabla P_1, \\ J_{m2} &= -\frac{M_2 D_{m12}}{RT}\nabla P_2 \end{aligned} \tag{6.14}$$

where, D_{m12} is the molecular diffusivity of methane and carbon dioxide, m^2/s. Molecular diffusion is a parameter for a two-component gas system, and its value is determined by both methane and carbon dioxide, so their diffusion coefficients in this system are the same, as follows:

$$D_{m12} = \frac{0.0101 \times T^{1.75} \times \sqrt{\left(\frac{1}{M_1} + \frac{1}{M_2}\right)} \times 10^{-3}}{P \times \left(\delta_1^{1/3} + \delta_2^{1/3}\right)^2} \tag{6.15}$$

where, δ_1 is methane molecular diffusion volume, cm^3/mol; δ_2 is carbon dioxide molecular diffusion volume, cm^3/mol. The mass flow flux caused by surface diffusion is as follows:

$$\begin{aligned} J_{s1} &= -\frac{D_{s1} C_{s1}}{P_1}\nabla P_1, \\ J_{s2} &= -\frac{D_{s2} C_{s2}}{P_2}\nabla P_2 \end{aligned} \tag{6.16}$$

where, C_{s1} is the methane adsorption capacity, kg/m^3; C_{s2} is the carbon dioxide adsorption capacity, kg/m^3. The calculation equation is as follows:

$$\begin{aligned} C_{s1} &= \frac{4\theta_1 M_1}{\pi d_1^3 N_A}, \\ C_{s2} &= \frac{4\theta_2 M_2}{\pi d_2^3 N_A} \end{aligned} \tag{6.17}$$

where, N_A is Avogadro constant, $6.022 \times 10^{23}/mol$. The surface diffusion coefficient (D_{s1} or D_{s2}) can be obtained by following equations:

$$D_{s1} = D_{s1}^0 \frac{(1-\theta_1) + \frac{\zeta}{2}\theta_1(2-\theta_1) + [H(1-\zeta)](1-\zeta)\frac{\zeta}{2}\theta_1^2}{\left(1-\theta_1+\frac{\zeta}{2}\theta_1\right)^2},$$

$$D_{s2} = D_{s2}^{0} \frac{(1-\theta_2) + \frac{\zeta}{2}\theta_2(2-\theta_2) + [H(1-\zeta)](1-\zeta)\frac{\zeta}{2}\theta_2^2}{\left(1-\theta_2+\frac{\zeta}{2}\theta_2\right)^2} \tag{6.18}$$

where,

$$H(1-\zeta) = \begin{cases} 0, & \zeta \geq 1 \\ 1, & 0 \leq \zeta \leq 1 \end{cases} \tag{6.19}$$

where, D_{s1}^{0} and D_{s2}^{0} are the initial surface diffusion coefficient for methane and carbon dioxide, m^2/s; ζ is the ratio of clogging rate constant to forward migration rate constant, dimensionless.

Therefore, the total gas mass flow equations of methane and carbon dioxide can be derived as follows:

$$J_1 = -\left\{ \frac{\phi\rho_1 r_t^2}{8\mu_1}(1+\alpha\cdot K_{n1})\left(1+\frac{4\cdot K_{n1}}{1-b\cdot K_{n1}}\right) + \frac{M_1 D_{k1}}{RT} + \frac{M_1 D_{m12}}{RT} + \frac{D_{s1}C_{s1}}{P_1} \right\}\nabla P_1 \tag{6.20}$$

$$J_2 = -\left\{ \frac{\phi\rho_2 r_t^2}{8\mu_2}(1+\alpha\cdot K_{n2})\left(1+\frac{4\cdot K_{n2}}{1-b\cdot K_{n2}}\right) + \frac{M_2 D_{k2}}{RT} + \frac{M_2 D_{m12}}{RT} + \frac{D_{s2}C_{s2}}{P_2} \right\}\nabla P_2 \tag{6.21}$$

The form of seepage velocity is as follows:

$$v_1 = \frac{J_1}{\rho_1} = -\left\{ \frac{\phi r_t^2}{8\mu_1}(1+\alpha\cdot K_{n1})\left(1+\frac{4\cdot K_{n1}}{1-b\cdot K_{n1}}\right) + \frac{D_{k1}}{P_1} + \frac{D_{m12}}{P_1} + \frac{ZRTD_{s1}C_{s1}}{M_1 P_1^2} \right\}\nabla P_1 \tag{6.22}$$

$$v_2 = -\left\{ \frac{\phi r_t^2}{8\mu_2}(1+\alpha\cdot K_{n2})\left(1+\frac{4\cdot K_{n2}}{1-b\cdot K_{n2}}\right) + \frac{D_{k2}}{P_2} + \frac{D_{m12}}{P_2} + \frac{ZRTD_{s2}C_{s2}}{M_2 P_2^2} \right\}\nabla P_2 \tag{6.23}$$

6.4 Gas Apparent Permeability Model

There is a mathematical relationship between gas apparent permeability and seepage velocity, which is knows as Darcy equation. The equation is as follows:

$$v = -\frac{k_{app}}{\mu}\nabla P \tag{6.24}$$

It can be obtained according to the assumptions:

$$v_1 = -x\left\{\frac{\phi r_t^2}{8\mu_1}(1+\alpha\cdot K_{n1})\left(1+\frac{4\cdot K_{n1}}{1-b\cdot K_{n1}}\right)+\frac{D_{k1}}{P_1}+\frac{D_{m12}}{P_1}+\frac{ZRTD_{s1}C_{s1}}{M_1P_1^2}\right\}\nabla P \tag{6.25}$$

$$v_2 = -y\left\{\frac{\phi r_t^2}{8\mu_2}(1+\alpha\cdot K_{n2})\left(1+\frac{4\cdot K_{n2}}{1-b\cdot K_{n2}}\right)+\frac{D_{k2}}{P_2}+\frac{D_{m12}}{P_2}+\frac{ZRTD_{s2}C_{s2}}{M_2P_2^2}\right\}\nabla P \tag{6.26}$$

The apparent permeability models of methane and carbon dioxide can be derived as follows:

$$k_{ai} = \frac{x\phi r_t^2}{8}(1+\alpha\cdot K_{n1})\left(1+\frac{4\cdot K_{n1}}{1-b\cdot K_{n1}}\right)+\frac{\mu_1 D_{k1}}{P}+\frac{\mu_1 D_{m12}}{P}+\frac{\mu_1 ZRTD_{s1}C_{s1}}{M_1PP_1} \tag{6.27}$$

$$k_{aj} = \frac{y\phi r_t^2}{8}(1+\alpha\cdot K_{n2})\left(1+\frac{4\cdot K_{n2}}{1-b\cdot K_{n2}}\right)+\frac{\mu_2 D_{k2}}{P}+\frac{\mu_2 D_{m12}}{P}+\frac{\mu_2 ZRTD_{s2}C_{s2}}{M_2PP_2} \tag{6.28}$$

6.5 Model Verification

The experimental data in Ref. [12] are compared with results from the simulation in this chapter. From Fig. 6.1, it is not difficult to find that the two results match very well, which verifies the correctness of the proposed model in this chapter.

6.6 Sensitive Parameters Analysis

Table 6.1 lists the parameters required to obtain the apparent permeability of methane and carbon dioxide. The influence factors of gas apparent permeability, contribution of flow mechanisms, and the influence of component ratio on apparent permeability are analyzed.

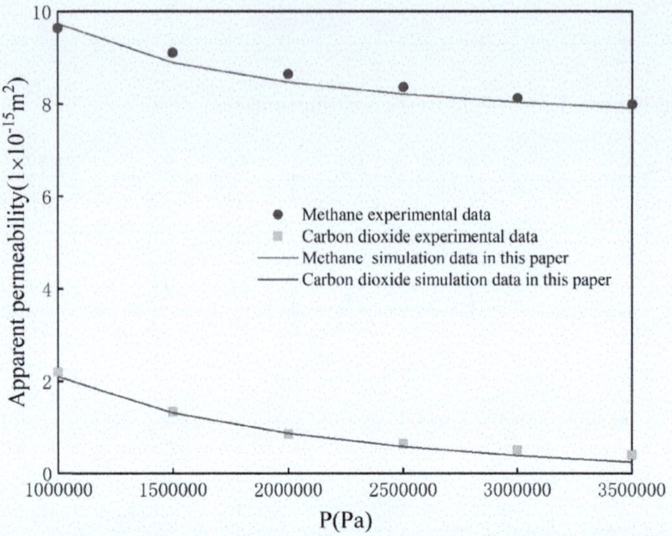

Fig. 6.1 Comparison between the experimental data and the simulated results in this chapter

Table 6.1 Parameters for solving the model

Parameters	Value	Meaning
d_1	0.4 nm	Methane molecular diameter
d_2	0.33 nm	Carbon dioxide molecular diameter
M_1	0.016 kg/mol	Molar mass of methane
M_2	0.044 kg/mol	Molar mass of carbon dioxide
k_b	1.38×10^{-23} J/K	Boltzmann constant
δ_1	20.1 cm^3/mol	Methane molecular diffusion volume
δ_2	24.9 cm^3/mol	Carbon dioxide molecular diffusion volume
R	8.314 J/mol/K	Ideal gas constant
P_{L1}	2.4 MPa	Langmuir pressure constant of methane
P_{L2}	1.6 MPa	Langmuir pressure constant of carbon dioxide
ϕ	0.05	Porosity
α_0	0.7543	The equation fitting constant
b	− 1	The fitting constant of slippage effect
N_A	6.02×10^{23} mol^{-1}	Avogadro constant
μ_1^0	0.011 mPa s	Initial viscosity of methane
μ_2^0	0.019 mPa s	Initial viscosity of carbon dioxide

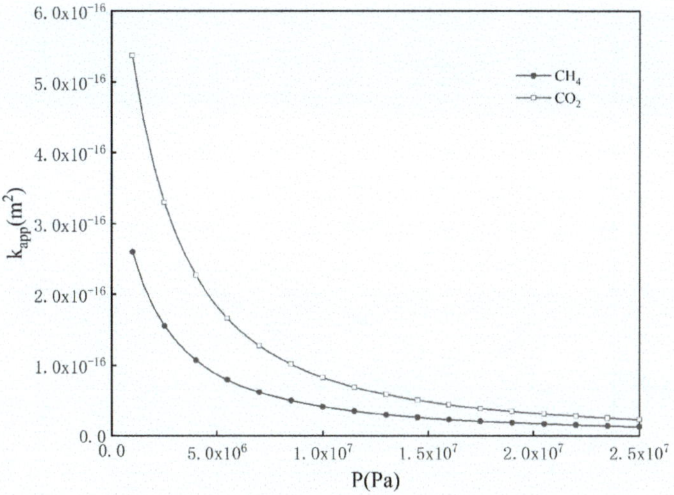

Fig. 6.2 Apparent permeability of methane and carbon dioxide varies with pressure ($r_m = 5$ nm)

6.6.1 *Influencing Factors of Apparent Permeability*

As can be seen from Fig. 6.2, pressure increasing leads to the decrease of apparent permeability for both CH_4 and CO_2. When the pressure increases from 1 to 5 MPa, the apparent permeability of the two-component gas changes obviously. However, when the pressure increases from 20 to 50 MPa, the apparent permeability curve basically does not change. This indicates that when the pressure exceeds 20 MPa, the effect of pressure on the gas apparent permeability curve can almost be ignored.

It can be seen from Fig. 6.3, pore radius increasing leads to the increase of apparent permeability for both methane and carbon dioxide. And the larger of the pore radius, the increasing trend increases more obvious. Comparing the apparent permeability of the two component gases, the apparent permeability of carbon dioxide is slightly higher than that of methane. At the same time, from the change of the permeability ratio, the ratio increases with pore radius. This shows that the change trend of methane apparent permeability is larger than that of carbon dioxide.

6.6.2 *Contribution of Flow Mechanisms to the Total Gas Transport*

The proportion of each diffusion mechanism is defined as follows:

$$\chi_{v1} = \frac{J_{v1}}{J_1},\quad \chi_{k1} = \frac{J_{k1}}{J_1},\quad \chi_{m1} = \frac{J_{m1}}{J_1},\quad \chi_{s1} = \frac{J_{s1}}{J_1} \tag{6.29}$$

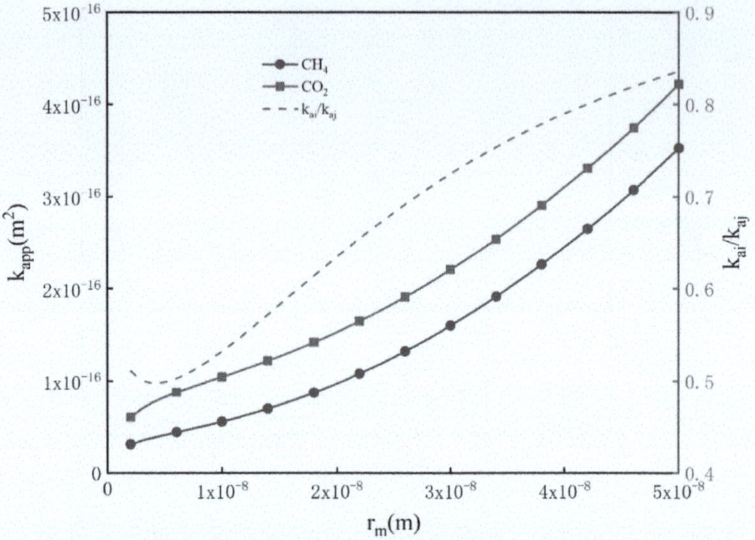

Fig. 6.3 The apparent permeability of methane and carbon dioxide and their ratio vary with pore radius ($P = 10$ MPa)

$$\chi_{v2} = \frac{J_{v2}}{J_2}, \chi_{k2} = \frac{J_{k2}}{J_2}, \chi_{m2} = \frac{J_{m2}}{J_2}, \chi_{s2} = \frac{J_{s2}}{J_2} \tag{6.30}$$

As can be seen from Figs. 6.4 and 6.5, the proportion (contribution) of flow mechanisms for methane is relatively uniform under low pressure. Among them, the proportion of slip flow and Knudsen diffusion increases with pore radius. However, the proportion of molecular diffusion and surface diffusion decreases with pore radius. Under low pressure, the main flow mechanism of carbon dioxide is surface diffusion. Among them, the proportion of slip flow and Knudsen diffusion increases with pore radius. And the proportion of molecular diffusion and surface diffusion decreases with pore radius. However, the decreasing trend of molecular diffusion is relatively less than that of surface diffusion.

Under high pressure, it can be seen from Figs. 6.6 and 6.7 that the main flow mechanisms for both methane and carbon dioxide are surface diffusion and slip flow. The proportion of surface diffusion decreases with pore radius. However, the proportion of viscous flow and Knudsen diffusion increases with pore radius. At the same time, the decreasing trend of surface diffusion for carbon dioxide is smaller than that of methane.

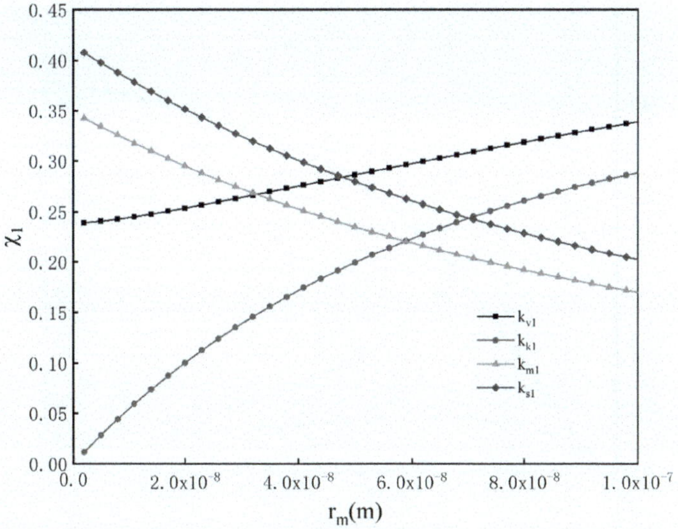

Fig. 6.4 The proportion of methane diffusion mechanism varies with pore radius ($P = 0.1$ MPa)

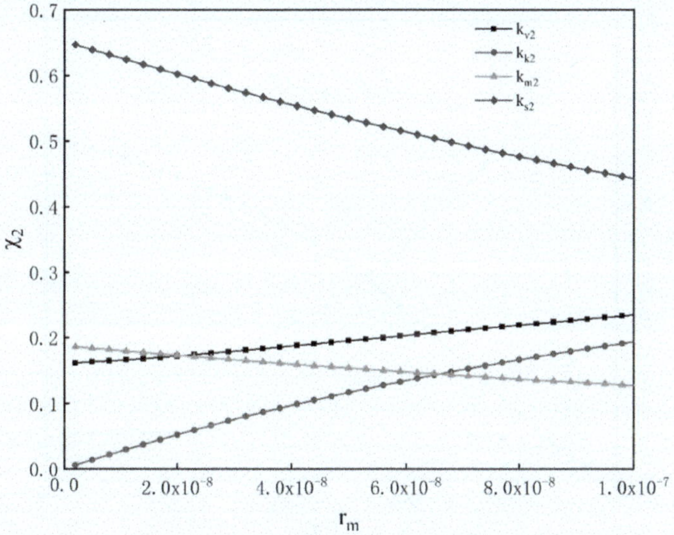

Fig. 6.5 The proportion of carbon dioxide diffusion mechanism varies with pore radius ($P = 0.1$ MPa)

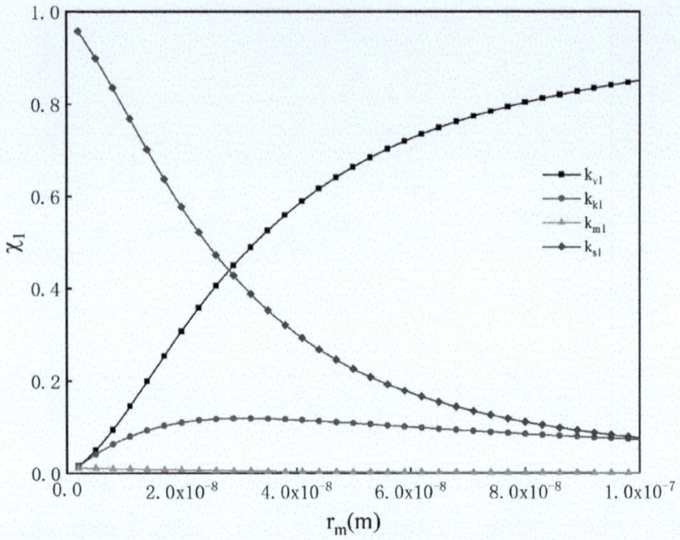

Fig. 6.6 The proportion of methane diffusion mechanism varies with pore radius ($P = 5$ MPa)

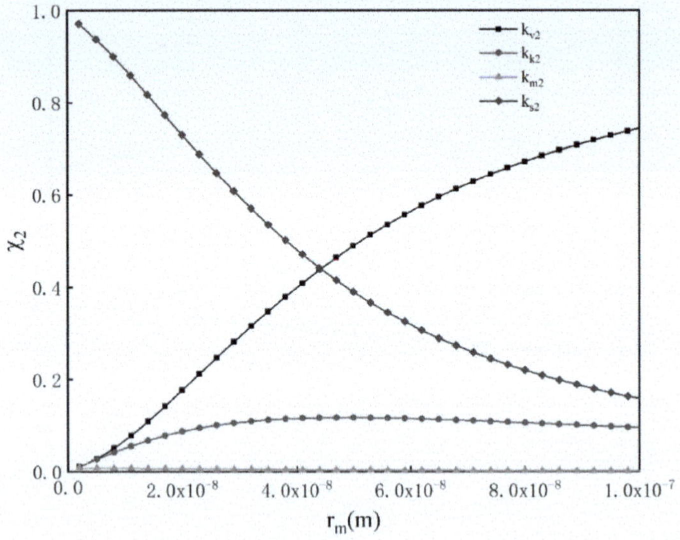

Fig. 6.7 The proportion of carbon dioxide diffusion mechanism varies with pore radius ($P =$ 5 MPa)

6.6.3 The Influence of Methane-Carbon Dioxide Component Ratio (MCCR)

The methane-carbon dioxide component ratio (MCCR) is defined as follows:

$$\begin{aligned}\varepsilon_{ij} &= \frac{x}{y} = \frac{x}{1-x},\\ \varepsilon_{ji} &= \frac{y}{x} = \frac{y}{1-y}\end{aligned} \tag{6.31}$$

The apparent permeability of the mixed gas is defined as follows:

$$v_g = v_1 + v_2 \tag{6.32}$$

$$v_g = \frac{k_{ag}}{\mu_g}\nabla P \tag{6.33}$$

By combining Eqs. (6.25) and (6.26), it can be obtained:

$$k_{ag} = \frac{\mu_g}{\mu_1}k_{ai} + \frac{\mu_g}{\mu_2}k_{aj} \tag{6.34}$$

As can be seen from Figs. 6.8, 6.9 and 6.10, the permeability of methane increases with MCCR, while the permeability of carbon dioxide decreases with MCCR. At the same time, the change trend of methane, carbon dioxide, and mixed gas permeability decreases with MCCR. At the same time, the relative change trend of methane and carbon dioxide permeability is less affected by pressure and pore radius. When the pressure and pore radius change, the relative change trend of methane and carbon dioxide permeability remains almost unchanged.

Under low pressure, the apparent permeability of mixed gas increases with MCCR. However, the trend changes with pressure increasing. The apparent permeability of mixed gas decreases with MCCR under high pressure. As the pressure increases, the apparent permeability of methane and carbon dioxide decreases. When the values of the two are close, the increase of methane is larger than that of carbon dioxide, which leads to the increase of the mixed gas the permeability. At the same time, the change of methane and carbon dioxide with MCCR is similar to that of methane and carbon dioxide with pressure. Therefore, the permeability of gas can also change only by adjusting MCCR even if pressure and other external conditions are constant.

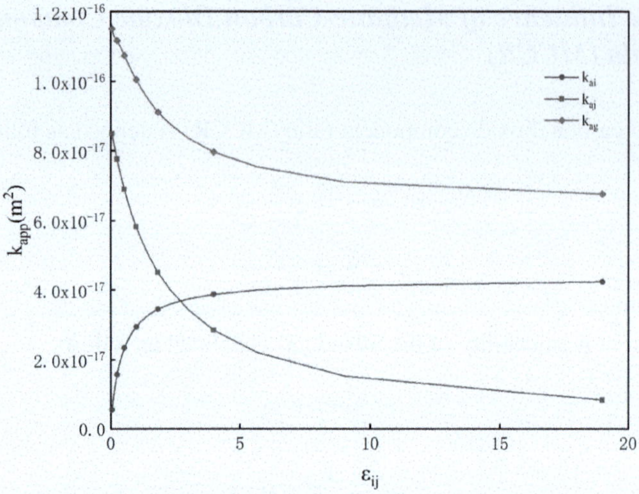

Fig. 6.8 The apparent permeability of methane, carbon dioxide and mixed gas, varies with component ratio ($P = 10$ MPa, $r_m = 5$ nm)

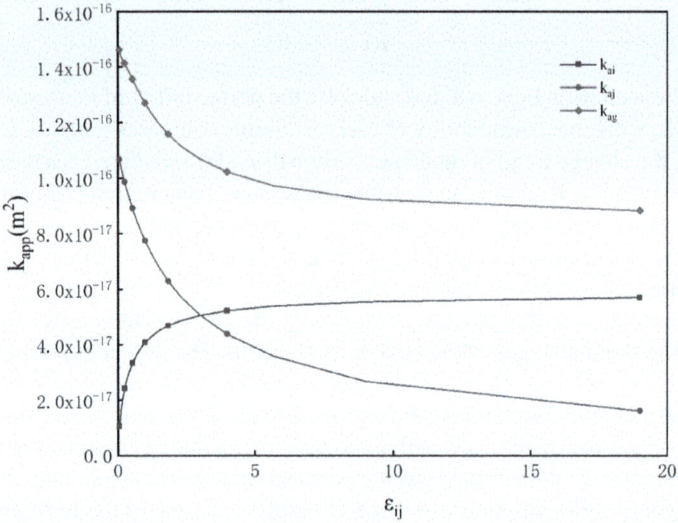

Fig. 6.9 The apparent permeability of methane, carbon dioxide and mixed gas, varies with component ratio ($P = 10$ MPa, $r_m = 10$ nm)

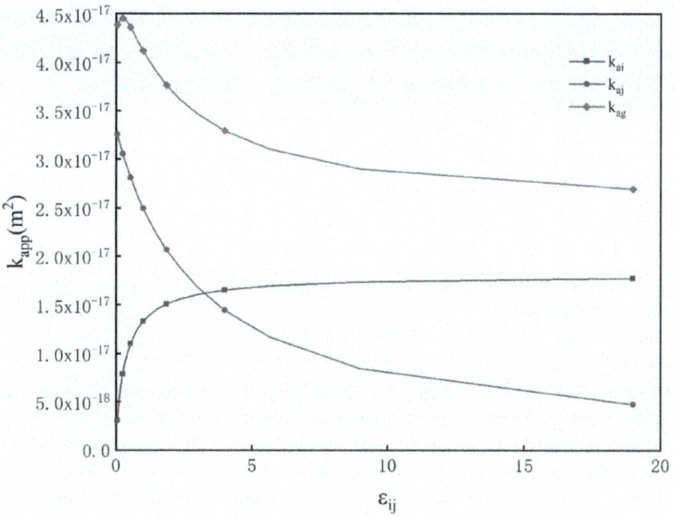

Fig. 6.10 The apparent permeability of methane, carbon dioxide and mixed gas, varies with component ratio ($P = 20$ MPa, $r_m = 5$ nm)

6.7 Remark

During the process of improving shale gas production by CO_2 injection, there are multiple and complex flow mechanisms co-exist in shale nano-pore, including viscous flow, molecular diffusion, Knudsen diffusion, and molecular diffusion. Above transport mechanisms can affect the apparent permeability of two-component gas. In this work, the apparent permeability of two-component gas has been studied to more accurately simulate this process. The main conclusions are as follows:

(1) Pressure increasing leads to the decrease of apparent permeability for both methane and carbon dioxide. When the pressure exceeds 20 MPa, the effect of pressure on the gas apparent permeability can almost be ignored. However, pore radius increasing leads to the increase of apparent permeability for both methane and carbon dioxide, and the change trend of methane apparent permeability is larger compared with that of carbon dioxide.
(2) The contribution (proportion) of slip flow and Knudsen diffusion to the total flow flux increases with pore radius increasing. However, the proportion of molecular diffusion and surface diffusion decreases with pore radius. Under high pressure, the main flow mechanisms for both methane and carbon dioxide are surface diffusion and slip flow. And the decreasing trend of surface diffusion contribution in carbon dioxide is smaller than that in methane.
(3) The change trend of methane and carbon dioxide permeability with MCCR is similar to pressure. The permeability of methane increases with MCCR, while the permeability of carbon dioxide decreases MCCR. The relative change trend

of methane and carbon dioxide permeability is less affected by pressure and pore radius. The permeability of gas can also change only by adjusting MCCR even if pressure and other external conditions are constant.

References

1. Lee, J., & Sidle, R. (2010). Gas-reserves estimation in resource plays. *SPE Economics & Management, 2*(02), 86–91.
2. Zhai, Z., et al. (2014). Adsorption and diffusion of shale gas reservoirs in modeled clay minerals at different geological depths. *Energy & Fuels, 28*(12), 7467–7473.
3. Zhang, H., & Cao, D. (2016). Molecular simulation of displacement of shale gas by carbon dioxide at different geological depths. *Chemical Engineering Science, 156*, 121–127.
4. Sui, H., & Yao, J. (2016). Effect of surface chemistry for CH_4/CO_2 adsorption in kerogen: A molecular simulation study. *Journal of Natural Gas Science and Engineering, 31*, 738–746.
5. Wu, H., Chen, J., & Liu, H. (2015). Molecular dynamics simulations about adsorption and displacement of methane in carbon nanochannels. *The Journal of Physical Chemistry C, 119*(24), 13652–13657.
6. Yuan, Q., et al. (2015). Molecular dynamics simulations of the enhanced recovery of confined methane with carbon dioxide. *Physical Chemistry Chemical Physics, 17*(47), 31887–31893.
7. Busch, A., et al. (2008). Carbon dioxide storage potential of shales. *International journal of greenhouse gas control, 2*(3), 297–308.
8. Ye, C., et al. (2013). Process analysis on CO_2 absorption by monoethanolamine solutions in microchannel reactors. *Chemical engineering journal, 225*, 120–127.
9. Hoffman, B. T., & Shoaib, S. (2014). CO_2 flooding to increase recovery for unconventional liquids-rich reservoirs. *Journal of Energy Resources Technology, 136*(2), 022801.
10. Sanchez-Rivera, D., Mohanty, K., & Balhoff, M. (2015). Reservoir simulation and optimization of Huff-and-Puff operations in the Bakken Shale. *Fuel, 147*, 82–94.
11. Iddphonce, R., & Wang, J. (2021). Investigation of CO_2 and CH_4 competitive adsorption during enhanced shale gas production. *Journal of Petroleum Science and Engineering, 205*, 108802.
12. Feng, C. (2017). Effects of CO_2 and CH_4 mixture binary gas on permeability of shale. *Journal of Hunan University of Science and Technology, 32*, 1–5.

Chapter 7
Two Phase Gas and Water Transport Model in Shale Gas Nanopores

Both gas and water transport exist in nanopores during the development of shale gas reservoirs (SGRs). The study of two-phase flow is conducive to the improvement of recovery for gas production. And it almost exists in the whole progression of SGR development. Based on the Beskok's model, Sun et al. built a new apparent permeability gas model with considering real gas effect [1], surface diffusion, stress dependence, the shape of different types of nanopores, and water film effect. Feng et al. studied the adsorption behavior of clay minerals in shale by measuring the adsorption isotherms of water [2]. Song et al. established water and gas migration models in shale nanopores, respectively [3]. The gas phase model in inorganic pores considers slippage effect, while the gas phase model in organic pores considers surface diffusion. Li et al. established two-phase gas–water relative permeability model with an interface effect grounded by the Poiseuille law [4], and verified the accuracy of the model. Shen et al. modeled the flow process of gas and water in hydraulic fracturing shale reservoirs through CMG software [5]. However, the flow mechanisms considered by the CMG software is not perfect due to the limitations of CMG, and it cannot follow up with the latest study in this area.

In this chapter, the relative permeability model is established by using porous medium fractal theory and tortuous capillary bundle model. Multiple gas transport mechanisms which revealed in recent years are included. Then, the model accuracy was proved by published data. Finally, based on the proposed model, sensitivity analysis was carried out, which include pressure, water saturation, slippage effect, pore throat structural parameters (tortuosity, pore size distribution), Knudsen and surface diffusions.

In SGRs, there are two main forms of shale gas exists which are bulk and adsorbed gases. The bulk gas exists in pores and fractures as free phase. The adsorption gas adsorbs on kerogen surface, clay surface, or pores. Correspondingly, there are three shale gas transport mechanisms which are shown as follows.

C. Guo and Z. Yang, *Shale Gas Production: Concept, Models, and Techniques*,
Petroleum Engineering, https://doi.org/10.1007/978-3-031-84552-9_7

7.1 Bulk Gas Transport in Nanopores

7.1.1 Viscous Slip Flow

In large pores or under high pressure, the gas mean free path is small. If $K_n < 10^{-3}$, the collision of gas molecules is dominated by intermolecular collisions. The collision of pore walls and gas molecules can be ignored. The gas flow conforms to the continuity condition and is the so called viscous flow. The distribution of the velocity of the fluid on the different diameter of pipe can be expressed by the continuity equation. The gas and water continuity equations are as follows:

Gas phase continuity equation:

$$\frac{\mu_g}{r_x}\frac{\partial}{\partial r_x}\left(r_x\frac{\partial v_g}{\partial r_x}\right) = \frac{\partial P_g}{\partial z} \tag{7.1}$$

Water phase continuity equation:

$$\frac{\mu_w}{r_x}\frac{\partial}{\partial r_x}\left(r_x\frac{\partial v_w}{\partial r_x}\right) = \frac{\partial P_w}{\partial z} \tag{7.2}$$

where v is velocity, with unit of m/s; r_x is the radial distance from any position in the capillary to the center of the capillary, m; μ_g and μ_w are the gas-phase and water-phase viscosities, respectively, Pa s; P_g and P_w are gas and water pressure in Pa; z is the length in the direction of flow, m. When the Knudsen number is $10^{-3} < \mathrm{K_n} < 10^{-1}$, interaction between molecules of gas and capillary wall can't be ignored, the gas molecules slip velocity on the pore wall is as follows:

$$v_{slip} = \left(C_1\lambda\frac{\partial v_g}{r_x} - C_2\lambda^2\frac{\partial^2 v_g}{\partial r_x^2}\right)\Bigg|_{r_x=r,} \tag{7.3}$$

where v_{slip} represent gas slip velocity on the wall, m/s; r stands for capillary radius, m; C_1 and C_2 are the first and second-orders coefficients of slip velocities; r_x is the diameter of the capillary center to any position directional distance, m; λ represents the gas mean free path:

$$\lambda = \frac{\mu_g}{P_g}\sqrt{\frac{\pi RT}{2M}} \tag{7.4}$$

where R stands for universal gas constant; T represents absolute temperature in K; M is molar mass of gas, kg/mol. The gas and water flows behavior in the nanotubes are shown in Fig. 7.1, and their boundary conditions in the capillaries can be expressed as:

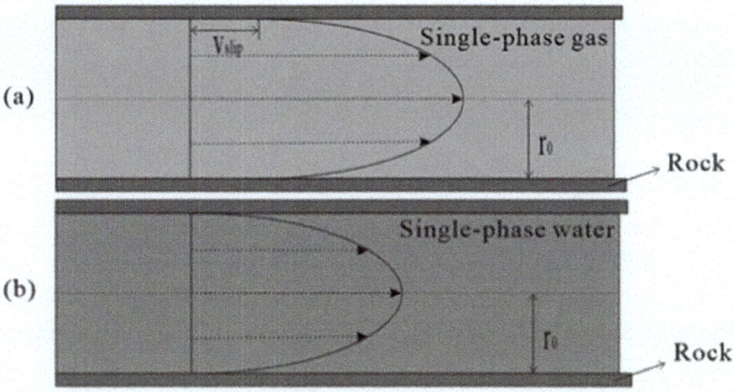

Fig. 7.1 Velocity distribution of single-phase gas (**a**) and water (**b**) in nanotubes. **a** The velocity distribution of gas molecules in the nanotube. **b** The velocity distribution of water molecules in the nanotube

$$\left.\frac{\partial v_g}{\partial r_x}\right|_{r_x=0} = 0; \quad v_g\big|_{r_x=r} = v_{slip} \tag{7.5}$$

$$\left.\frac{\partial v_w}{\partial r_x}\right|_{r_x=0} = 0; \quad v_w|_{r_x=r} = 0 \tag{7.6}$$

According to Eqs. (7.1)–(7.4) and boundary conditions (7.5)–(7.6), the velocity distribution in the capillary can be obtained:

$$v_{vs} = -\frac{\Delta P}{L}\frac{1}{4\mu_g}r_x^2 + \frac{\Delta P}{L}\frac{1}{4\mu_g}\left(r^2 + 2C_1\lambda r - 2C_2\lambda^2\right) \tag{7.7}$$

$$v_w = -\frac{\Delta P}{L}\frac{1}{4\mu_w}r_x^2 + \frac{\Delta P}{L}\frac{1}{4\mu_w}r^2 \tag{7.8}$$

where v_{vs} represents viscous slip flow velocity, m/s; v_w stands for the water flow velocity, m/s; L represents pore path, m; ΔP represents the pressure difference along the transport path, Pa. By integrating the Eqs. (7.7)–(7.8) with respect to the pipe diameter r_x along the capillary tube, the volumetric flow rate is as follows:

$$q_{vs} = \int_0^r v_{vs} dA_r = \frac{\Delta P}{L}\frac{\pi}{8\mu_g}\left(r^4 + 4C_1\lambda r^3 - 4C_2\lambda^2 r^2\right) \tag{7.9}$$

$$q_w = \int_0^r v_w dA_r = \frac{\Delta P}{L}\frac{\pi}{8\mu_w}r^4 \tag{7.10}$$

Here q_{vs} stands for volumetric flow rate of gas viscous slip flow, m^3/s, and q_w is the volumetric flow rate of water phase, m^3/s.

7.1.2 *Knudsen Diffusion*

When the pore size or the gas pressure is small, the gas mean free path of gas molecules becomes large. If the Knudsen number $K_n \gg 1$. The gas molecules mostly collide with the pore wall. The mass flow of Knudsen diffusion is as follows:

$$J_k = -MD_k \frac{\Delta C}{L} \tag{7.11}$$

$$D_k = \frac{2}{3} r \sqrt{\frac{8RT}{\pi M}} \tag{7.12}$$

$$\rho = \frac{PM}{RT}, \; C = \frac{\rho}{M} = \frac{P}{RT} \tag{7.13}$$

where J_k represents mass flow rate produced by Knudsen diffusion, kg/m^2 s; D_k represents Knudsen diffusion coefficient of methane molecule, m^2 s; C stands for gas molar concentration of methane molecule, mol/m^3; L is the pore path, m; ρ is density of methane, kg/m^3; P stands for gas pressure, Pa. When gas molecules flow in Knudsen diffusion mode, gas diffusion is more sensitive to wall roughness. When the wall roughness increases, the gas stays longer around the pore wall surface and the Knudsen diffusion efficiency decreases. The mass flow of gas, as a result of Knudsen diffusion process considering the roughness of the wall, can be expressed as follows:

$$J_k = -\frac{2}{3} r \sigma^{D_s - 2} \sqrt{\frac{8M}{\pi RT}} \frac{\Delta P}{L} \tag{7.14}$$

where ΔP is the pressure difference on the transport path L, Pa; D_s represents fractal measurement of the pore wall surface, unitless; σ represents ration of molecular to local pore diameters, unitless. Therefore, the volume flux of Knudsen diffusion is as follows:

$$q_k = -\frac{J_k}{\rho} A = \frac{2}{3} \frac{\pi r^3}{P} \sigma^{D_s - 2} \sqrt{\frac{8RT}{\pi M}} \frac{\Delta P}{L} \tag{7.15}$$

where q_k is the volume flux caused by Knudsen diffusion, m^3/s; A is the capillary cross-sectional area, m^2.

7.1.3 *Weight Calculation of Viscous Slip Flow and Knudsen Diffusion*

During SGR production process, Knudsen number K_n of the gas transport of the nanopore is 0.0002–6. Intermolecular and molecule-pore wall collisions are equally important and cannot be ignored. Gas transmission is mainly slippage flow and transitional flow, partly continuous flow. The number of collisions of gas molecules per unit time in the nanopore can be expressed as:

$$\frac{1}{t_T} = \frac{1}{t_M} + \frac{1}{t_S} \tag{7.16}$$

$$\frac{1}{t_T} = \frac{\overline{v}}{\lambda_T} \tag{7.17}$$

$$\frac{1}{t_M} = \frac{\overline{v}}{\lambda} \tag{7.18}$$

$$\frac{1}{t_S} = \frac{\overline{v}}{2r} \tag{7.19}$$

where t_T stands for regular time required for a collision of molecules, s; t_M stands for time required for molecules to collides between molecules, s; t_S represents regular collisions time required for molecule and the pore wall, s; λ_T represents mean free path of gas molecules, m; $\overline{v}$ represents mean molecular velocity, m/s.

$$\frac{\overline{v}}{\lambda_T} = \frac{\overline{v}}{\lambda} + \frac{\overline{v}}{2r} \tag{7.20}$$

$$\frac{1}{t_M} : \frac{1}{t_S} = \frac{2r}{\lambda} = \frac{1}{K_n} \tag{7.21}$$

The total transport volume of nanopore bulk gas is the weight sum of viscous slip flow and Knudsen diffusion:

$$q_b = \frac{1}{1 + K_n} q_{vs} + \frac{K_n}{1 + K_n} q_k \tag{7.22}$$

$$K_n = \frac{\lambda}{d} \tag{7.23}$$

where q_b stands for volume flow rate of bulk gas in nanopores, m^3/s. d is the pore diameter, m.

7.1.4 Stress Sensitive

Stress sensitivity of shale is very strong because the organic matter strength is weak. The stress change will cause the change of pore diameter and transport capacity of shale organic nanopores. This change in permeability due to stress sensitivity can be expressed as follows:

$$\omega_m = \left[1 + \frac{(\alpha_f - \alpha_s)}{(1 + \overline{K}_n s/E_s)} \frac{s(P - P_{int})}{b_{int} E_s}\right] \tag{7.24}$$

where ω_m is the stress-sensitive permeability coefficient of shale organic matter, dimensionless; α_f is the micro-crack Biot coefficient, dimensionless; α_s is the shale organic matter Biot coefficient, dimensionless; $\overline{K}_n$ is the micro-crack stiffness, Pa/m; s is the spacing of microcracks, m; E_s is the Young's modulus of organic matter, Pa; b_{int} is the initial opening of microcracks, m; P_{int} is the initial pressure, Pa.

7.1.5 Matrix Contraction

In shale nanopores, the existence of adsorbed gas takes portion of the nanopore bulk gas transmission space and reduces the gas transmission capacity. During the development of depressurization, the adsorbed gas desorbs. Therefore, the effective hydrodynamic diameter and transmission capacity of the bulk gas increase in this process, which is the so called shrinkage effect. This change in permeability due to the shrinkage effect of organic matter can be expressed as follows:

$$\omega_s = \left[1 - \frac{3}{\phi} \frac{\varepsilon_L p_L (P - P_{int})}{(P + P_L)(P_{int} + P_L)}\right]^3 \tag{7.25}$$

where ω_s is the permeability coefficient due to shale organic matter shrinkage effect, dimensionless; ε_L is the Langmuir strain, dimensionless; p_L is the Langumir pressure, Pa; ϕ is the porosity.

7.2 Coupling of Gas Transport Mechanisms in Nanopores

In this chapter, multiple transport mechanisms have been comprehensively considered, including the effects of slippage effect, Knudsen diffusion, pore roughness, stress sensitivity and matrix shrinkage effect. By combining Eqs. (7.22) to (7.25), the volume flow rate of the bulk gas transmission in nanopore can be expressed as:

$$q_{bt} = \xi_{mb} \omega_m \omega_s \cdot q_b \tag{7.26}$$

where q_{bt} is the corrected volume flow rate of the bulk gas transmission in nanopore, m^3/s; ξ_{mb} is the correction factor for bulk gas transport in nano pores of SGRs.

Under the initial conditions in SGRs, the adsorbed gas and bulk gas are in equilibrium. The amount of adsorption gas can be obtained by the Eq. 7.27.

$$q_a = \frac{q_L P}{P_L + P} \tag{7.27}$$

where q_a is the adsorption volume, m^3/kg; P is pressure, Pa; q_L is Langmuir volume, m^3/kg. Since Langmuir adsorption is a monolayer adsorption, coverage of adsorbed gas (θ) can be obtained by through the ratio of the amount adsorbed q_a to the Langmuir volume q_L, namely:

$$\theta = \frac{P}{P_L + P} \tag{7.28}$$

where θ is the observed amount of the gas under equilibrium condition. The concentration of the single-layer adsorbed gas can be obtained from Eq. (7.28):

$$C_s = \frac{4\theta M}{\pi d_M^3 N_A} \tag{7.29}$$

where C_s stands for gas concentration, kg/m^3; N_A is Avogadro's constant, $N_A = 6.0221415 \times 10^{23}\ mol^{-1}$; d_M is the molecular diameter, m; M represents the molecular weight of methane, kg/mol. With reference to Maxwell–Stefan equation [6], amount of diffused gas is:

$$J_s = -\xi_{ms} L_m C_s \frac{\Delta\mu}{L} \tag{7.30}$$

where J_s represents mass flux due to surface diffusion, kg/m^2 s; ξ_{mb} stands for correction coefficient for shale surface diffusion, dimensionless; L_m represents mobility, mol s/kg; μ is the partial molar free energy, J/mol; L is the transport path, m. Assuming the gas is under ideal state condition, the expression of the partial molar free energy (chemical potential) can be obtained:

$$\mu = \mu_o + RT \ln P \tag{7.31}$$

where μ_o is the partial molar free energy of the reference state, J/mol; P is pressure, Pa. Compared with rate of adsorption and desorption, surface diffusion is very small. By combining Eqs. (7.30) and (7.31), the amount of diffused gas on the pore surface can be further derived as follows:

$$J_s = -\xi_{ms} D_s^0 \frac{C_s}{P} \frac{\Delta P}{L} \tag{7.32}$$

where D_s^0 represents diffusion constants on the surface at the time adsorbed gas volume is zero, m^2/s. Hwang's model is commonly used to characterize the surface diffusion coefficient [6]:

$$D_s^0 = \Omega T^m \exp\left(-\frac{E}{RT}\right) \tag{7.33}$$

where Ω represents values related to molar mass, m^2/(s K$^{0.5}$); E is activation energy of gas, J/mol; m is constant, dimensionless, which can be expressed as follows:

$$E = \Delta H^{0.8} \tag{7.34}$$

Combining Eqs. (7.33) and (7.34), the surface diffusion coefficient can be calculated as follows:

$$D_s^0 = 8.29 \times 10^{-7} T^{0.5} \exp\left(-\frac{\Delta H^{0.8}}{RT}\right) \tag{7.35}$$

It should be noted that D_s^0 in Eqs. (7.33) and (7.35) are not related to pressure [7] and the calculated value is far from the actual value under high pressure. With considering effect of pressure, D_s can be further represented as follows:

$$D_s = D_s^0 \frac{(1-\theta) + \frac{\kappa}{2}\theta(2-\theta) + [H(1-\kappa)](1-\kappa)\frac{\kappa}{2}\theta^2}{\left(1-\theta+\frac{\kappa}{2}\theta^2\right)} \tag{7.36}$$

$$H(1-\kappa) = \begin{cases} 0, \ \kappa \geq 1 \\ 1, \ 0 \leq \kappa \leq 1 \end{cases}, \quad \kappa = \frac{\kappa_b}{\kappa_m} \tag{7.37}$$

where D_s stands for coefficients of surface diffusion, m^2 s; θ is the surface coverage, unitless; κ is the coefficient representing blocking surface gas, dimensionless; $H(1-\kappa)$ stands for Heaviside function, dimensionless; κ_m represents surface gas molecule advancing velocity constant, m/s; κ_b is the surface gas molecule blocking velocity constant, m/s. Combining the above equations, the mass flow of surface diffusion is:

$$\begin{aligned} J_s &= -\xi_{ms} D_s \frac{C_s}{P}\frac{\Delta p}{L} \\ &= -\frac{(1-\theta) + \frac{\kappa}{2}\theta(2-\theta) + [H(1-\kappa)](1-\kappa)\frac{\kappa}{2}\theta^2}{\left(1-\theta+\frac{\kappa}{2}\theta^2\right)} \\ &\quad \times 8.29 \times 10^{-7} T^{0.5} \exp\left(-\frac{\Delta H^{0.8}}{RT}\right) \frac{4M}{\pi d_m^3 N_A} \frac{1}{P_{L+}P} \frac{\Delta P}{L} \end{aligned} \tag{7.38}$$

Then, the surface diffusion volume flux is:

$$q_s = -J_s\frac{\pi r^2}{\rho} = \xi_{ms}D_s\frac{C_s RT\pi r^2}{P^2 M}\frac{\Delta P}{L} \tag{7.39}$$

Combining Eqs. (7.9), (7.15), (7.22), (7.25), and (7.38), the volume flux of gas and water in the capillaries can be obtained:

$$\begin{aligned} q_g =& \frac{\xi_{mb}\omega_m\omega_s}{1+K_n}\frac{\pi}{8\mu_g}\frac{\Delta P}{L}\left(r^4 + 4C_1\lambda r^3 - 4C_2\lambda^2 r^2\right) \\ &+ \frac{\xi_{mb}\omega_m\omega_s}{1+1/K_n}\frac{2r_0^3\sigma^{D_s-2}}{3P}\sqrt{\frac{8\pi RT}{M}}\frac{\Delta P}{L} \\ &+ \xi_{ms}D_s\frac{C_s RT\pi r^2}{P^2 M}\frac{\Delta P}{L} \end{aligned} \tag{7.40}$$

$$q_w = \frac{\nabla P}{L}\frac{\pi}{8\mu_w}r^4 \tag{7.41}$$

where Q_g is the volume flux of gas, m^3/s; J_w is the volume flux of water, m^3/s.

7.3 Establishment of Gas–Water Relative Permeability Model

7.3.1 *Fractal Theory*

Regardless of the macro-scale or micro-scale, the characteristics of fractal self-similar are ubiquitous in porous media. According to the fractal geometry theory, the cumulative pore size distribution is presented as follows:

$$N(\varepsilon > r) = \left(\frac{r_{\max}}{r}\right)^{D_f} \tag{7.42}$$

where N is the pore's number; ε is the dimensional length, m; r and $r_{\max}$ are the pore size for a special pore and maximum pore size, m; D_f is the fractal dimension, dimensionless, and the range of two-dimensional and three-dimensional pores, respectively is $1 < D_f < 2$ and $2 < D_f < 3$. D_f can be found by through the following equation:

$$D_f = d - \frac{\ln\phi}{\ln(r_{\min}/r_{\max})} \tag{7.43}$$

where $d = 2$ or 3 known as Euclidean space, dimensionless; $r_{\min}$ is the minimum size of the pore, m; ϕ is porosity. The pore's number from minimum size to the maximum size can be obtained as follows:

$$N = \left(\frac{r_{\max}}{r_{\min}}\right)^{D_f} \tag{7.44}$$

The probability distribution density of the pores having unalike pore sizes is given as a function as follows:

$$f(r) = D_f r_{\min}^{D_f} r^{-(D_f+1)} \tag{7.45}$$

Since the actual pore path is tortuous, the continuous characterization unit (REV) of the porous medium with a characteristic length of L_0 is taken. The actual length L is larger than L_0, and L is calculated by through following equation:

$$L(r) = (2r)^{1-D_T} L_0^{D_T} \tag{7.46}$$

where D_T is the fractal dimension. For 2D, $1 < D_T < 2$; For 3D, $2 < D_T < 3$. When $D_T = 1$ imply straight capillary, $D_T = 2$ (two Dimension) or 3 (three-dimensional), it means that the capillary tube is infinitely tortuous and can fill the entire space. It can be expressed as:

$$D_T = 1 + \frac{\ln \tau_{ave}}{\ln(L_0/2r_{ave})} \tag{7.47}$$

$$\tau_{ave} = \frac{1}{2} + \frac{1}{4}\sqrt{1-\phi} + \frac{1}{2}\sqrt{1-\phi}\frac{\sqrt{\left(\frac{1}{\sqrt{\sqrt{1-\phi}}} - 1\right)^2 + \frac{1}{4}}}{1-\sqrt{1-\phi}} \tag{7.48}$$

$$\frac{L_0}{2r_{ave}} = \frac{D_f - 1}{D_f^{1/2}}\left[\frac{1-\phi}{\phi}\frac{\pi}{4(2-D_f)}\right]^{1/2}\frac{r_{\max}}{r_{\min}} \tag{7.49}$$

7.3.2 Derivation of Gas–Water Relative Permeability Model

The capillary bundle model can be used to characterize the flow path of fluid in heterogeneous shale, which includes a series of tortuous capillaries of unequal diameters. The direction of the capillary bundle is perpendicular to the representation unit, and no fluid flow outside the capillary bundle. As revealed in Fig. 7.2, the pore size distribution and tortuosity of the capillary bundle model conform to the fractal theory introduced by Eqs. (7.42)–(7.48). The capillary force in the small pores in the shale is large, and it is difficult for the gas to enter. Therefore, there exists a critical pore radius r_c. When pore radius is less than r_c, only water exists. When $r > r_c$, the center is occupied by gas and there is a water film near the pore wall. And it should be noted that the smooth-thin wetting film can be ignored compared to gas due to

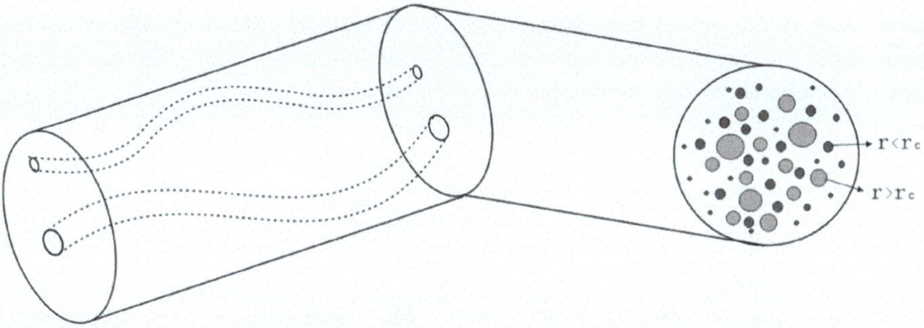

Fig. 7.2 Illustration of gas and water distribution. The capillary bundle consist of a large number of tortuous capillary tubes of different diameters (where the capillary tubes with radius greater than r_c are filled with gas, and the capillary tubes with radius less than r_c are filled with water)

its extremely thin thickness. With reference to above assumptions, we can obtain the formula for calculating water saturation:

$$S_w = \frac{\int_{r_{\min}}^{r_c} \pi r^2 f(r) dr}{\int_{r_{\min}}^{r_{\max}} \pi r^2 f(r) dr} = \frac{r_c^{-D_f+2} - r_{\min}^{-D_f+2}}{r_{\max}^{-D_f+2} - r_{\min}^{-D_f+2}} \tag{7.50}$$

The value of D_f is as follows:

$$D_f = d - \frac{\ln \phi}{\ln(r_{\min}/r_{\max})} \tag{7.51}$$

And the value of r_c is as follows:

$$r_c = r_{\max}[(1-\phi)S_w + \phi]^{\frac{1}{-D_f+2}} \tag{7.52}$$

By introducing the tortuosity and the pressure gradient of each phase, the volumetric flow equations (7.40) and (7.41) of gas and water in one single tube can be expressed as:

$$\begin{aligned} q_g = {} & \frac{\xi_{mb}\omega_m\omega_s}{1+K_n}\frac{\Delta P_g}{L(r)}\frac{\pi}{8\mu_g}\left(r^4 + 4C_1\lambda r^3 - 4C_2\lambda^2 r^2\right) \\ & + \frac{\xi_{mb}\omega_m\omega_s}{1+1/K_n}\frac{2r^3\sigma^{D_s-2}}{3P}\sqrt{\frac{8\pi RT}{M}}\frac{\Delta P_g}{L(r)} \\ & + \xi_{mb}D_s\frac{C_s RT\pi r^2}{P^2 M}\frac{\Delta P_g}{L(r)} \end{aligned} \tag{7.53}$$

$$q_w = \frac{\Delta P_w}{L(r)}\frac{\pi}{8\mu_g}r^4 \tag{7.54}$$

where $L(r)$ is the actual path length; $\Delta P_g/L(r)$ and $\Delta P_w/L(r)$ are the pressures gradients of the gas phase and water phase. By integrating Eq. (7.53), the gas volume flow of the characterization unit (REV) can be obtained:

$$
\begin{aligned}
Q_g &= N\int_{r_c}^{r_{\max}} q_g f(r)dr = \frac{\xi_{mb}\omega_m\omega_s}{1+\overline{K_n}}\frac{\pi}{8\mu_g}\frac{D_f r_{\max}^{D_f}\Delta P_g}{2^{1-D_T}L_0^{D_T}}\cdot\\
&\times\left(\frac{r_{\max}^{D_T-D_f+3}-r_c^{D_T-D_f+3}}{D_T-D_f+3}+\frac{r_{\max}^{D_T-D_f+2}-r_c^{D_T-D_f+2}}{D_T-D_f+2}4C_1\lambda\right.\\
&\left.-\frac{r_{\max}^{D_T-D_f+1}-r_c^{D_T-D_f+1}}{D_T-D_f+1}4C_1\lambda^2\right)\\
&+\frac{\xi_{mb}\omega_m\omega_s}{1+1/K_n}\frac{2\sigma^{D_s-2}}{3P}\sqrt{\frac{8\pi RT}{M}}\frac{D_f r_{\max}^{D_f}\Delta P_g}{2^{1-D_T}L_0^{D_T}}\frac{r_{\max}^{D_T-D_f+2}-r_c^{D_T-D_f+2}}{D_T-D_f+2}\\
&+\xi_{ms}\frac{\pi D_s RTC_S}{P^2M}\frac{D_f r_{\max}^{D_f}\Delta P_g}{2^{1-D_T}L_0^{D_T}}\frac{r_{\max}^{D_T-D_f+1}-r_c^{D_T-D_f+1}}{D_T-D_f+1}
\end{aligned}
\tag{7.55}
$$

Similarly, by integrating the Eq. (7.54) from the capillary radius r_c to $r_{\max}$, we can obtain the water volume flow of the characterization unit (REV):

$$
\begin{aligned}
Q_w &= N\int_{r_{\min}}^{r_c} q_g f(r)dr\\
&= N\int_{r_{\min}}^{r_c}\frac{\Delta P_w}{L(r)}\frac{\pi}{8\mu_w}r^4 D_f r_{\min}^{D_f} r^{-(D_f+1)}dr\\
&= \frac{\pi}{8\mu_w}\frac{D_f r_{\max}^{D_f}\Delta p_w}{2^{1-D_T}L_0^{D_T}}\frac{r_c^{D_T-D_f+3}-r_{\min}^{D_T-D_f+3}}{D_T-D_f+3}
\end{aligned}
\tag{7.56}
$$

In addition, the flow rates of the gas and water can be presented by the Darcy equations:

$$
Q_g = -\frac{K_g}{\mu_g}\frac{A}{(1-S_w)}\frac{\Delta P_g}{L_0}
\tag{7.57}
$$

$$
Q_w = -\frac{K_w}{\mu_w}\frac{A}{S_w}\frac{\Delta P_w}{L_0}
\tag{7.58}
$$

where K represent the apparent permeability;A is cross sectional area of the characterization unit (REV). The apparent permeability can be obtained by combining Eqs. (7.55) to (7.58):

$$
\begin{aligned}
K_g &= -(1-S_w)\frac{\mu_g L_0}{\Delta P_g A}\cdot Q_g \\
&= -(1-S_w)\frac{\mu_g}{A}\frac{\xi_{mb}\omega_m\omega_s}{1+\overline{K_n}}\frac{\pi}{8\mu_g}\frac{D_f r_{\max}^{D_f}}{2^{1-D_T}L_0^{D_{T-1}}} \\
&\times\left(\frac{r_{\max}^{D_T-D_f+3}-r_c^{D_T-D_f+3}}{D_T-D_f+3}+\frac{r_{\max}^{D_T-D_f+2}-r_c^{D_T-D_f+2}}{D_T-D_f+2}4C_1\lambda\right. \\
&\left.-\frac{r_{\max}^{D_T-D_f+1}-r_c^{D_T-D_f+1}}{D_T-D_f+1}4C_1\lambda^2\right) \\
&-(1-S_w)\frac{\mu_g}{A}\frac{\xi_{mb}\omega_m\omega_s}{1+1/\overline{K_n}}\frac{2\sigma^{D_s-2}}{3P}\sqrt{\frac{8\pi RT}{M}}\frac{D_f r_{\max}^{D_f}}{2^{1-D_T}L_0^{D_{T-1}}}\left.\frac{r_{\max}^{D_T-D_f+2}-r_c^{D_T-D_f+2}}{D_T-D_f+2}\right] \\
&-(1-S_w)\frac{\mu_g}{A}\xi_{ms}\frac{\pi D_s RTC_S}{P^2M}\frac{D_f r_{\max}^{D_f}}{2^{1-D_T}L_0^{D_{T-1}}}\frac{r_{\max}^{D_T-D_f+1}-r_c^{D_T-D_f+1}}{D_T-D_f+1}
\end{aligned}
\tag{7.59}
$$

$$
\begin{aligned}
K_w &= -S_w\frac{\mu_w L_0}{\Delta p_w A}\cdot Q_w \\
&= -S_w\frac{\mu_w}{A}r_{\max}^{D_f}\frac{D_f}{2^{1-D_T}L_0^{D_T-1}}\frac{\pi}{8\mu_w}\frac{r_c^{D_T-D_f+3}-r_{\min}^{D_T-D_f+3}}{D_T-D_f+3}
\end{aligned}
\tag{7.60}
$$

where $\overline{K_n}$ is the equivalent Knudsen number of the capillary bundle under pressure P. When Eq. (7.60) satisfies the condition of $S_w = 1$, the absolute permeability of the core can be obtained:

$$
K = -\frac{1}{A}\frac{\pi}{8}\frac{D_f r_{\max}^{D_f}}{2^{1-D_T}L_0^{D_T-1}}\frac{r_{\max}^{D_T-D_f+3}-r_{\min}^{D_T-D_f+3}}{D_T-D_f+3}
\tag{7.61}
$$

Therefore, the K_{rg} and K_{rw} in shale reservoirs can be well-defined as follows:

$$
\begin{aligned}
K_{rg} &= (1-S_w)\frac{\xi_{mb}\omega_m\omega_s}{1+\overline{K_n}}\left\{1-\left(\frac{r_c}{r_{\max}}\right)^{D_T-D_f+3}\right. \\
&+\frac{D_T-D_f+3}{D_T-D_f+2}\left[\frac{1}{r_{\max}}-\frac{1}{r_c}\left(\frac{r_c}{r_{\max}}\right)^{D_T-D_f+3}\right]4C_1\lambda \\
&\left.-\frac{D_T-D_f+3}{D_T-D_f+1}\left[\frac{1}{r_{\max}^2}-\frac{1}{r_c^2}\left(\frac{r_c}{r_{\max}}\right)^{D_T-D_f+3}\right]4C_2\lambda^2\right\} \\
&+(1-S_w)\frac{\xi_{mb}\omega_m\omega_s}{1+1/\overline{K_n}}\frac{16\sigma^{D_s-2}\mu_g}{3P} \\
&\sqrt{\frac{8RT}{\pi M}}\frac{D_T-D_f+3}{D_T-D_f+2}\left[\frac{1}{r_{\max}}-\frac{1}{r_c}\left(\frac{r_c}{r_{\max}}\right)^{D_T-D_f+3}\right]
\end{aligned}
$$

$$+ (1 - S_w)\xi_{ms}\frac{8\mu_g D_s RTC_S}{P^2 M}\frac{D_T - D_f + 3}{D_T - D_f + 1}\left[\frac{1}{r_{\max}^2} - \frac{1}{r_c^2}\left(\frac{r_c}{r_{\max}}\right)^{D_T - D_f + 3}\right] \tag{7.62}$$

$$K_{rw} = S_w\left(\frac{r_c}{r_{\max}}\right)^{D_T - D_f + 3} \tag{7.63}$$

7.3.3 Model Verification

The gas–water relative permeability simulation data (Table 7.1) is used to verify the proposed model. In order to make the established model consistent with the actual situation, the normalized water saturation $S_{w0} = (S_w - S_{w0})/(1 - S_{w0})$ is used for verification [8]. The curves of relative permeability of the model herein and those of Piquemal [9] and Zhang et al. [10] are shown in Fig. 7.3. It is observed that the established model herein can obtain a good match with Piquemal's result. Also, compared with Zhang's model, the proposed model fully considered multiple gas transport mechanisms. Therefore, there is some difference in the gas relative permeability between the proposed model and Zhang's model.

7.4 Sensitive Parameters' Analysis

7.4.1 Pressure and Water Saturation

The K_{rg} and K_{rw} at different pressures is shown in Fig. 7.4. The K_{rg} decreases with pressure increasing. The effect of pressure on K_{rg} becomes less significant under high pressure. The reason is when pressure is low, Knudsen diffusion, slippage effect, and surface diffusions all contribute to the gas transport. With pressure increasing, the gas flow properties tend to liquid, and these flow effects are gradually weakened.

Also, from Fig. 7.4 above, as S_w increasing, the K_{rg} decreases which correlate directly with many laboratory experiments results. This is because the increase of S_w in SGRs always starts from small pores. That is, the increase of water saturation, those small pores tend to be occupied by water instead of the gas. Thus, the volume of the gas starts to decrease, and finally the relative permeability of the gas tends to decrease.

Table 7.1 Summary of calculation parameters

Parameter description	Symbol	Unit	Value
Shale gas viscosity	μ_g	Pa s	1.75×10^{-5}
Simulated pressure P	P	MPa	0.15
Simulated temperature	T	K	423.15
Methane molecular diameter	d_c	m	4.14×10^{-9}
First-order slip coefficient	C_1	–	1/3
Second-order slip coefficient	C_2	–	0.25
Euclidean space	d	–	2
Shale average porosity	φ	–	0.05
Gas constant	R	J/(mol K)	8.314462
Gas molar mass	M	kg/mol	0.016
Molecular diameter to local pore diameter ration	σ	–	0.5
Microcrack Biot coefficient	α_f	–	0.8
Shale organic matter Biot coefficient	α_s	–	0.65
Microcrack stiffness	$\overline{K}_n$	Pa m^{-1}	3.58×10^{12}
Micro crack spacing	s	m	2.50×10^{-2}
Organic Young's Modulus	E_s	Pa	4.00×10^{9}
Micro crack initial opening	b_{int}	m	5.00×10^{-6}
Langmuir strain	ε_L	–	3.00×10^{-3}
Langmuir pressure	P_L	Pa	6.72×10^{6}
Langmuir volume	q_L	m^3/kg	0.05
Isobaric heat of adsorption when the gas coverage is "0"	ΔH	J/mol	1.4×10^{4}
Gas wall diffusion blocking coefficient	κ	–	0.5

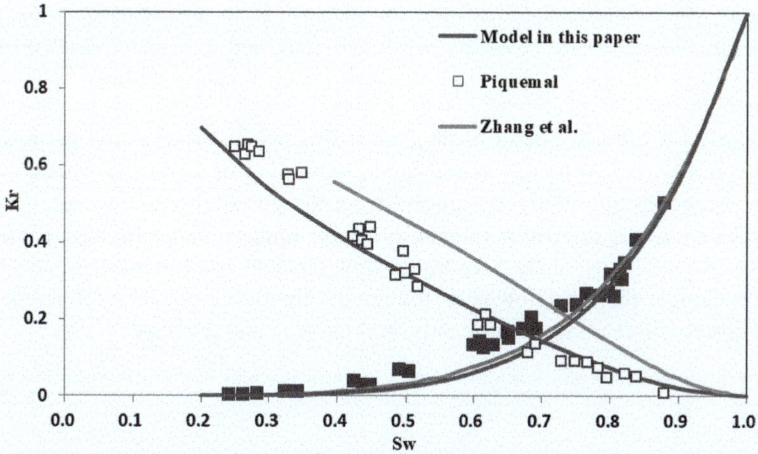

Fig. 7.3 Comparison of the proposed model against existing models

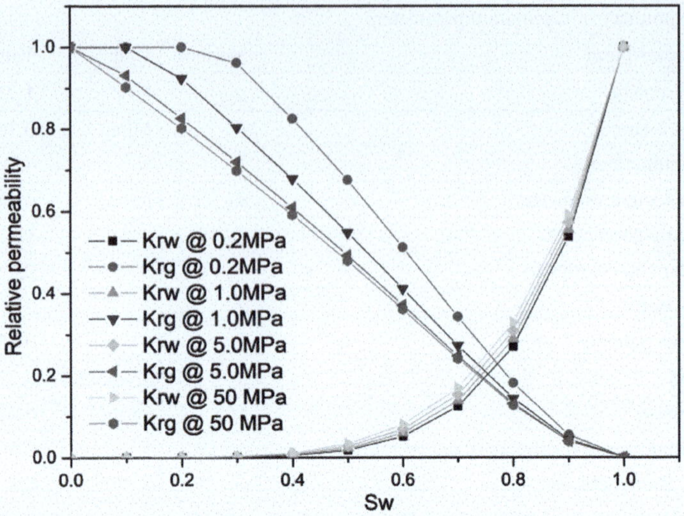

Fig. 7.4 Relative permeability of gas and water at different pressures

7.4.2 *Pore Structure Parameters*

Pore structure parameters includes pore size distribution fractal dimension (D_f) and tortuosity fractal dimension (D_T). The effect of D_f is shown in Fig. 7.5. According to Fig. 7.5, when water saturation is the same, increasing of pore size distribution fractal dimension D_f improves the K_{rg}. However, it is not conducive to K_{rw} because it tends to decrease. This is because, the higher of pore size distribution fractal dimension indicates that the critical pore radius r_c corresponding to the same water saturation will be reduced. Therefore, the gas will flow in relatively larger pores and the gas phase will be relatively infiltrated. The rate will increase accordingly. Similarly, because water flows more in small pores, greater capillary force makes K_{rw} decrease.

Similarly, the influence of tortuosity fractal dimension D_T on relative permeability is shown in Fig. 7.6. When S_w is the same, increasing of tortuosity fractal dimension D_T improves the relative permeability of the gas slightly. However, it is also not conducive to the relative permeability of the aqueous phase because it tends to decrease. As we know, shale pore-throat flow channel has a high degree of tortuosity. The longer the fluid flow path, the greater the flow resistance. Thus, the rock permeability will decrease significantly, and the K_{rw} will decrease.

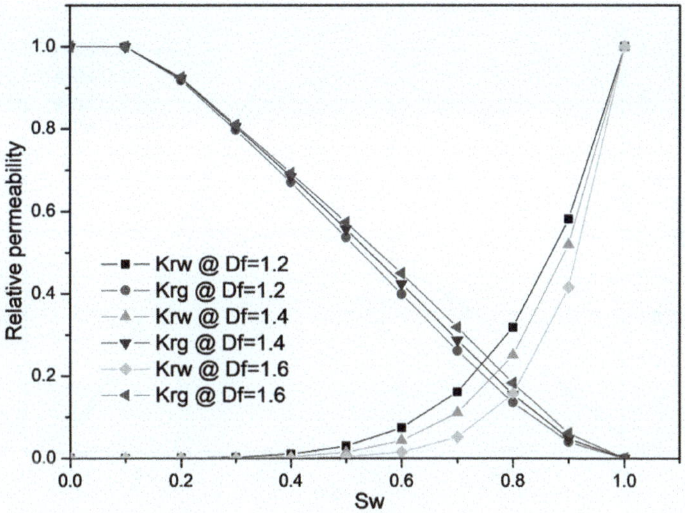

Fig. 7.5 The effect of pore size distribution fractal dimension D_f on relative permeability

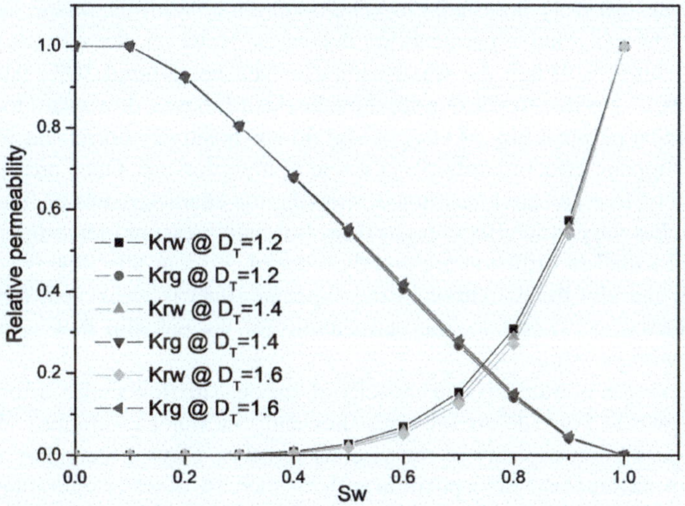

Fig. 7.6 The influence of tortuosity fractal dimension D_T on relative permeability

7.4.3 Effect of Viscous Slip Flow

The relative permeability curve due to viscous slip flow under different pressures is shown in Fig. 7.7. According to the Fig. 7.7a, when the pressure is low, the relative permeability of gas due to viscous slip flow increases with K_{rw} increasing. When S_w is in low stage, r_c is positively correlated with water saturation. Knudsen and surface

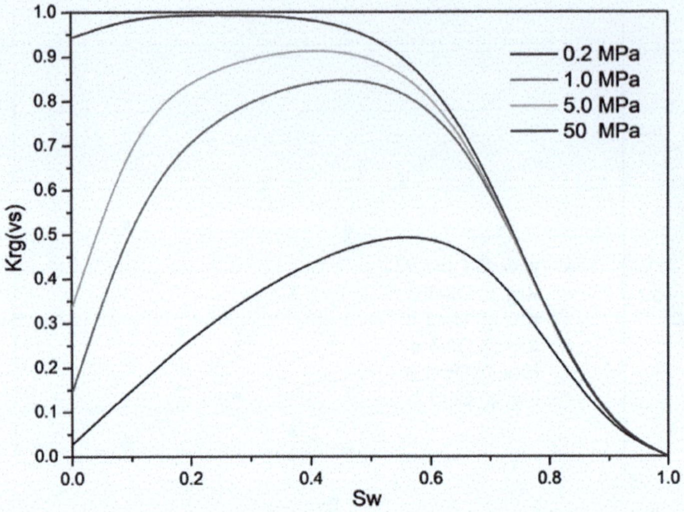

Fig. 7.7 Relative permeability curve of viscous slip flow under different pressures

diffusions are inversely correlated to water saturation. With the water saturations increasing, the gas relative permeability decreases. However, the viscous slip flow still rises relatively. When the water saturation reaches around 50%, the relative permeability of gas due to viscous slip flow begins to show a downward trend.

The relative permeability of viscous slip flow is positively correlated with pressure. The slippage effect mainly plays a role at low pressure. Once pressure rises, the degrees of free gas are constrained, showing the characteristics of liquid. Slippage effect has very little effect on gas flow. But with pressure increasing, effect of surface and Knudsen diffusion are greatly reduced. Viscous slip flow is composed by viscous and slip flows. Although the slippage effect is weakened, the viscous slip flow increases. Therefore, the permeability of viscous slip flow is positively correlated with pressure.

The proportion of transmission capacity of viscous slip flow under different pressures is shown in Fig. 7.8. At low pressures and low water saturations, the transmission contribution capacity of the viscous slip flow plays a secondary role, and gradually increases to the main position with the increase of water saturation or pressure. Also, as it can be found from Fig. 7.8, at low pressure and low water saturation, the transmission contribution capacity of viscous slip flow is of secondary importance, and gradually increases to the main position with increase of water saturation or increase of pressure.

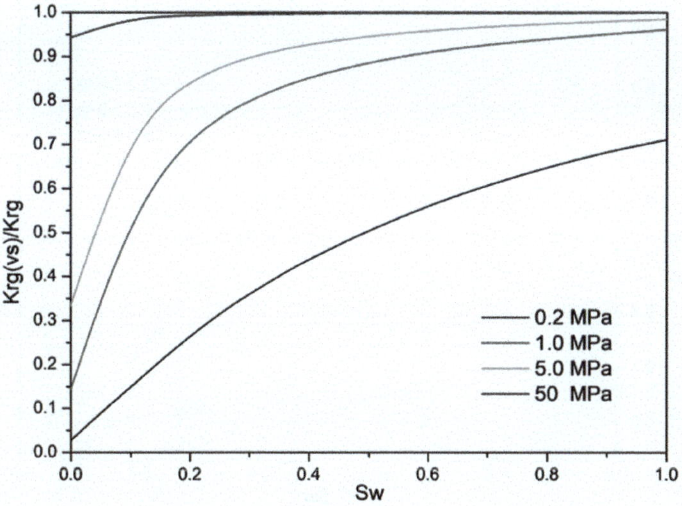

Fig. 7.8 Proportion of transmission capacity of viscous slip flow under different pressures

7.4.4 *The Impact of Knudsen Diffusion*

Similarly, the relative permeability curve due to Knudsen diffusion and the contribution of Knudsen diffusion versus pressure are analyzed (Figs. 7.9 and 7.10). In small pores and under low pressure, the average molecular gas free path λ becomes larger than the pore diameter. The collisions of gas molecules with the pore wall will dominate. And the bulk gas transport is mainly contributed by Knudsen diffusion (Fig. 7.9). According to the Fig. 7.9, at a pressure of 0.2 MPa and a water content of 0–10%, the relative permeability of Knudsen diffusion has increased slightly. The relative permeability of Knudsen diffusion shows a more apparent decrease as the pressure rises. This is due to the gas free path λ and the weight coefficient of Knudsen diffusion are inversely related to the pressure. From 0.2 to 1 MPa, the relative permeability of Knudsen diffusion drops sharply.

According to Fig. 7.10, when the pressure is below 0.2 MPa and the water content is less than 20%, the transmission contribution capacity of Knudsen diffusion is more than 50%. Under condition of $P > 1$ MPa and $S_w > 40\%$, Knudsen diffusion is negligible.

7.4.5 *Effect of Surface Diffusion*

Furthermore, the relative permeability curve due to surface diffusion and the proportion of transmission capacity of surface diffusion under different pressures are shown in Figs. 7.11 and 7.12, separately. According to Fig. 7.11, the K_{rg} due to surface

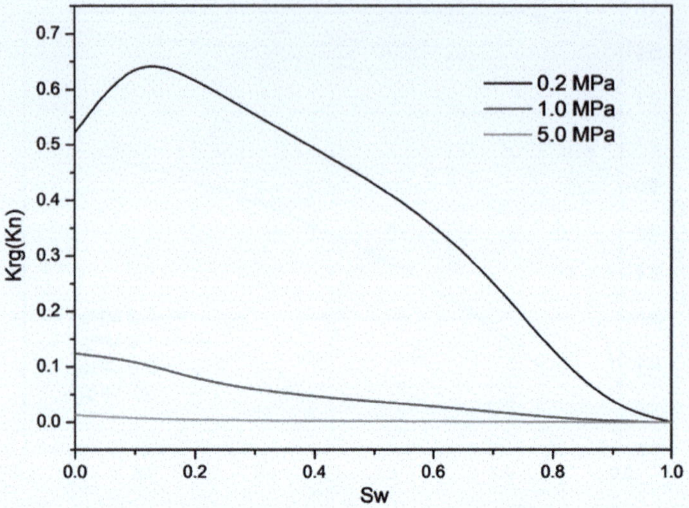

Fig. 7.9 Relative permeability curve of Knudsen

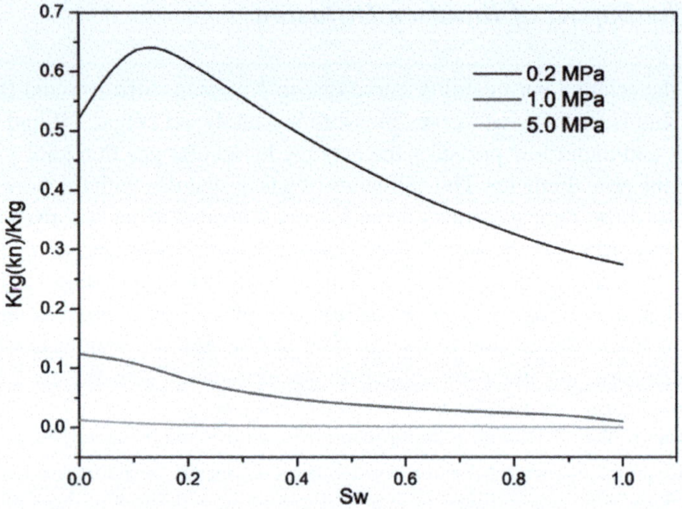

Fig. 7.10 Proportion of transmission capacity of Knudsen

diffusion is inversely proportional to the water saturation. When the water content is 0–20%, it decreases rapidly, and when the water content is greater than 20%, it declines gently. This is because as the water saturation rises, small pores are gradually occupied by water, r_c gradually increases. The changing tendency of the relative permeability due to surface diffusion vs. pressure is: When $P < 1$ MPa, it is proportional to the pressure. When $P > 1$ MPa, it shows oppositional trend. The reason is:

the surface diffusion is related to C_s/P^2 in addition to the surface diffusion coefficient D_s. And C_s/P^2 is negatively related with P, while D_s is positively related with P. When $P < 1$ MPa, the D_s is increasing with pressure. When $P > 1$ MPa, with pressure increasing, the decrease of C_s/P^2 is larger than the increase of D_s.

As shown in the Fig. 7.12, when the pressure is below 1 MPa and water saturation is less than 10%, the surface diffusion contribution is 18.30–73.24%. Once the pressure

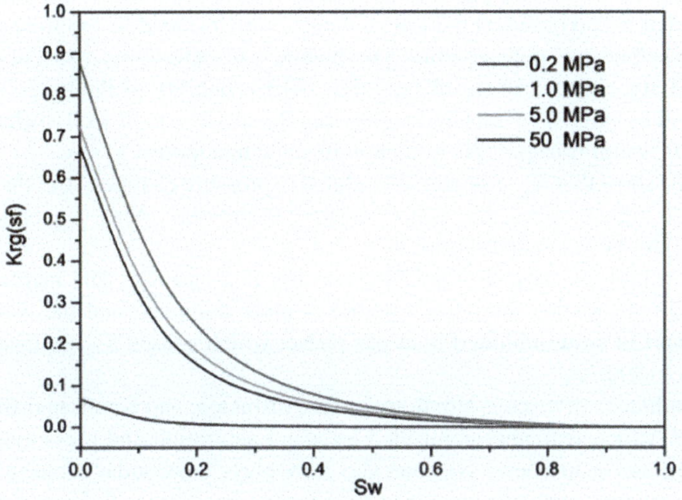

Fig. 7.11 Relative permeability curve of surface diffusion under different pressures

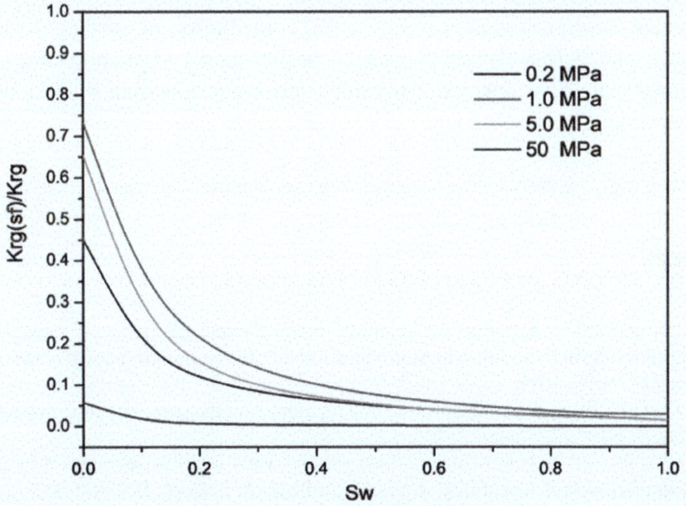

Fig. 7.12 Proportion of transmission capacity of surface diffusion under different pressures

is kept above 1 MPa and saturation of water is greater than 70%, the surface diffusion contribution rate is less than 4.54%, which can be ignored.

7.5 Remark

(1) In this chapter, the two-phase gas–water transmission model in shale nanopores was established based on the fractal theory, which considers multiple gas transport mechanisms, pore structure parameters, water saturation distribution, stress sensitivity, and matrix shrinkage effect. The reliability of the model has been verified by published data, indicating that the model can fit well to characterize the gas–water transport phenomenon in shale nanopores.
(2) Results show that K_{rg} is negatively related to pressure change. After the pressure rises to higher values, K_{rg} is not sensitive to pressure. Also, K_{rg} is negatively related to the water saturation.
(3) The D_f and D_T have similar effects on the K_{rw} and K_{rg}. The increase of D_f and D_T means that the pore structure is more complex, and the water phase transport is more inhibited than gas phase, which causes K_{rg}. to increase and K_{rw} to decrease.
(4) The influence of viscous slip flow, surface diffusion, and Knudsen diffusion are mixed, which is mainly controlled by water saturation and pressure. When P and S_w are in low level, viscous slip flow plays a secondary role in the total transport, and gradually increases to the main position as the water saturation or pressure increases.
(5) When pressure is kept above 1 MPa and water saturation is greater than 40%, the role of Knudsen diffusion is negligible. When pressure is kept below 1 MPa and water saturation is kept below 10%, contribution of surface diffusion can reach 21.64–72.78%. When pressure is higher than 1 MPa and water saturation is greater than 70%, surface diffusion contributes less than 4.25%, which can be overlooked.

References

1. Sun, Z., et al. (2017). Apparent permeability model for real gas transport through shale gas reservoirs considering water distribution characteristic. *International Journal of Heat and Mass Transfer, 115*, 1008–1019.
2. Feng, D., et al. (2018). Water adsorption and its impact on the pore structure characteristics of shale clay. *Applied Clay Science, 155*, 126–138.
3. Song, W., et al. (2019). Nanoscale confined gas and water multiphase transport in nanoporous shale with dual surface wettability. *Advances in Water Resources, 130*, 300–313.
4. Li, T., et al. (2016). An analytical method for modeling and analysis gas-water relative permeability in nanoscale pores with interfacial effects. *International Journal of Coal Geology, 159*, 71–81.

5. Shen, W., et al. (2016). Numerical simulation of gas and water flow mechanism in hydraulically fractured shale gas reservoirs. *Journal of Natural Gas Science and Engineering, 35*, 726–735.
6. Hwang, S. T., & Kammermeyer, K. (1966). Surface diffusion in microporous media. *The Canadian Journal of Chemical Engineering, 44*(2), 82–89.
7. Xu, P., et al. (2013). Prediction of relative permeability in unsaturated porous media with a fractal approach. *International Journal of Heat and Mass Transfer, 64*, 829–837.
8. Piquemal, J. (1994). Saturated steam relative permeabilities of unconsolidated porous media. *Transport in Porous Media, 17*, 105–120.
9. Zhang, T., et al. (2018). A fractal model for gas–water relative permeability in inorganic shale with nanoscale pores. *Transport in Porous Media, 122*, 305–331.
10. Krishna, R., and Wesselingh J. (1997). The Maxwel Stefan approach to mass transfer. Chem Eng Sci 52 (6): 861–911. https://doi.org/10.1016/S0009-2509(96)00458-7.

Chapter 8
Multi-stage Horizontal Well Hydraulic Fracturing Technic and Model for Shale Gas Production

Unconventional tight shale reservoirs have been developed during recent years due to increasing shortage of conventional resources and a large number of multi-stage fractured horizontal wells (MsFHW) have been drilled to enhance reservoir production performance. Gas flow in tight shale reservoirs is a multi-mechanism process, including: desorption, diffusion, and non-Darcy flow. The productivity of the shale gas reservoir with MsFHW is influenced by both reservoir condition and hydraulic fracture properties. In this chapter, a dual porosity model was constructed to estimate the effect of parameters on shale gas production with MsFHW. The simulation model was verified with the available field data from the Barnett Shale. Following flow mechanisms have been considered in this model: viscous flow, slip flow, Knudsen diffusion, and gas desorption. Langmuir isotherm was used to simulate the gas desorption process. Sensitivity analysis on production performance of tight shale reservoirs with MsFHW have been conducted. Parameters influencing shale gas production were classified into two categories: reservoir properties including matrix permeability, matrix porosity and hydraulic fracture properties including hydraulic fracture spacing, fracture half-length. Typical ranges of matrix parameters have been reviewed. Sensitivity analysis have been conducted to analyze the effect of above factors on the production performance of shale gas reservoirs. Through comparison, it can be found that hydraulic fracture parameters are more sensitive compared with reservoir parameters. And reservoirs parameters mainly affect the later production period. However, the hydraulic fracture parameters have significant effect on gas production from the early period. Result of this study can be used to improve the efficiency of history matching process. Also, it can contribute to the design and optimization of hydraulic fracture treatment design in unconventional tight reservoirs.

C. Guo and Z. Yang, *Shale Gas Production: Concept, Models, and Techniques*,
Petroleum Engineering, https://doi.org/10.1007/978-3-031-84552-9_8

8.1 Models for Simulating Fluid Flow in Fractured Reservoirs

Many different models have been proposed to simulate fluid flow in fractured reservoirs. These methods include: (1) equivalent continuum model; (2) dual porosity model, including dual porosity single permeability model, and dual porosity dual permeability model; (3) Multiple Interaction Continua (MINC) method; (4) Multiple porosity model; (5) Discrete fracture model. Though the discrete fracture model is more rigorous than the first three models, it is still limited in applications to field study due to computational intensity and lack of detailed knowledge of fracture distribution.

Equivalent continuum model. In the equivalent continuum model, the fractured porous media is considered as a continuum media. The properties of the fracture arrays, such as direction, location, permeability, porosity, etc., are averaged into the whole porous media. This method has not been widely used in shale gas reservoir simulation as it is not easy to obtain the averaged permeability tensor and it is too conceptual. Moridis et al. [1] have constructed effective continuum shale gas reservoir simulation model with considering muti-component gas adsorption. In the effective continuum model, the shale gas reservoir is discretized and the fractures are characterized with grid cells as single plannar planes or network of planar planes.

Dual Porosity Model. Dual porosity model is widely used to simulate the fluid flow in fractured porous media and many scholars have published a lot of literature based on dual porosity model. There are two interacting media: the matrix blocks with high porosity and low permeability, and the fracture network with low porosity and high conductivity. According to whether there is fluid flow in the matrix, the dual porosity model can be classified into dual porosity single permeability model and the dual porosity dual permeability model. In the dual porosity single permeability model, the flow only occur through the fracture system and matrix are treated as the spatially distributed sinks or sources to the fracture system. Watson et al. [2] have performed production data analysis for Devonian shale gas reservoir using DPM and presented analytical reservoir production models for history matching and production forecasting. Bustin et al. [3] have constructed a 2D dual porosity model to simulated shale gas reservoirs which uses both experimental and field data as model input parameters. Ozkan et al. [4] have developed a dual-mechanism dual-porosity model to simulate the linear flow for fractured horizontal well in shale gas reservoirs. In the dual porosity dual permeability model, the flow in the matrix is considered additionally. Connell and Lu have proposed an algorithm for the dual-porosity model considering Darcy flow in the fractures and diffusion in the matrix [5]. Moridis et al. [1] constructed a dual permeability model and compared it with dual porosity model and effective continuum model. Results showed that dual permeability model offered the best production performance. Renyi et al. [6] have proposed a new simulation model for shale gas reservoir based on dual porosity dual permeability model with considering multiple flow mechanisms, such as adsorption/desorption, diffusion, viscous flow, and deformation.

Multiple Interaction Continua method. As an extension for dual porosity model, Pruess [7] have proposed a multiple interaction continua method (MINC) to more accurately characterize the interaction between matrix and fracture. The matrix system in every grid block for MINC method will be subdivided into a sequence of nested rings which will make it possible to more accurately calculate the interblock fluid flow. In comparison with dual porosity model, MINC method is able to describe the gradient of pressures, temperatures, or concentrations near matrix surface and inside the matrix which can provide a better numerical approximation for the transient fracture matrix interaction [8]. Wu et al. [8] have proposed a generalized mathematical model to simulate multiphase flow in tight fractured reservoirs using MINC method with considering following mechanisms: Klinkenberg effect, non-Newtonian behavior, non-Darcy flow, and rock deformation.

Multiple Porosity Model. Some scholars also proposed multiple porosity model in which matrix is further separated into two or three parts with different properties. Civan et al. [9] proposed a qual-porosity model to simulate shale gas reservoirs, including organic matter, inorganic matter, natural, and hydraulic fractures. Dehghanpour and Shirdel [10] have proposed a triple porosity model to simulate shale gas reservoirs by incorporating micro fractures into the dual porosity models. They assumed that the matrix blocks in dual porosity model is composed of submatrix with nano Darcy permeability, and micro fractures with milli to micro Darcy permeability. Yan et al. [11] presented a micro-scale model. In Yan's model, the shale matrix was classified into inorganic matrix and organic matrix. The organic part was further divided into two parts basing on pore size of kerogen: organic materials with vugs and organic materials with nanopores. Therefore, there are four different continua in this model: nano organic matrix, vug organic matrix, inorganic matrix, and micro fractures.

Discrete Fracture Model. In the discrete fracture model, the fractures are discretely modeled. Each fracture is characterized as a geometrically well-defined entity. In the modeling of shale gas reservoirs using discrete fracture model, gas flow from the matrix to the discrete fractures by diffusion and desorption. Mayerhofer et al. [12] have investigated the fluid flow in fractured shale gas reservoirs considering stimulated reservoir volume (SRV) using explicit fracture network [12]. Gong et al. [13] have proposed a systematic methodology of building discrete fracture model based on microseismic interpretations for shale reservoirs with considering multiple mechanisms, such as gas adsorption, non-Darcy effect, etc.

The discrete fracture model is more rigorous, however, application of this method to reservoir scale is limited due to computational intensity and lack of detailed knowledge of fracture matrix properties and spatial distribution in reservoirs. And the dual porosity model can deal with the fracture matrix interaction more easily and less computationally demanding than discrete fracture method. So, dual porosity model is the main approach for modeling fluid and heat flow in fractured porous media.

8.2 Fluid Flow Mechanisms in Tight Shale Reservoirs

Due to the complexity of flow in shale gas reservoirs, three flow mechanisms should be considered in the dual-permeability model: viscous flow, Knudsen diffusion, slip flow, and gas adsorption.

8.2.1 *Viscous Flow*

This is the region where Knudsen number (The ratio of gas mean free path to pore throat diameter) has fairly low values such as in the conventional reservoirs with pore diameter in micrometers. In this Knudsen number range, the free path of the gas is smaller than the pore diameter (Knudsen number is far less than 1), so the motion of the gas molecules is mainly affected by the collision between molecules. The molecule's collision with the pore walls are negligible. The gas flow is driven by the pressure gradient between the single component gas, which can be characterized using the Darcy law:

$$J_v = -\frac{\rho_m k_{mi}}{\mu_g}(\nabla p_m) \tag{8.1}$$

where J_v is the mass flow (kg/(m^2 s)), ρ_m is gas density in bedrock (kg/m^3), k_{mi} is the intrinsic permeability of bedrock (m^2), μ_g is the gas viscosity (Pa s), p_m is pore pressure in the bedrock (Pa).

8.2.2 *Knudsen Diffusion*

This is the region where the Knudsen number is larger than 1 where the pore diameter is small enough so that the mean free path of the gas is close to the pore diameter. Under this Knudsen range, the gas flow is dominated by the collision between the gas molecules and the wall surface. The gas mass flow can be characterized using the Knudsen diffusion equation [14] which is shown in Eq. (8.2):

$$J_k = -M_g D_{km}(\nabla C_m) \tag{8.2}$$

where J_k is the mass flow caused by Knudsen diffusion (kg/(m^2 s)), M_g is the gas molar mass (kg/mol), D_{km} is the diffusion coefficient of the bedrock (m^2/s), C_m is the gas mole concentration (mol/m^3).

Table 8.1 Knudsen's permeability correction factor for tight porous media

Model	Correlation factor
Klinkenberg	$b_k = 4c\lambda\overline{P}/r$
Heid et al	$b_k = 11.419(k_\infty)^{-0.39}$
Jones and Owens	$b_k = 12.639(k_\infty)^{-0.33}$
Sampath and Keighin	$b_k = 13.851(k_\infty/\phi)^{-0.53}$
Florence et al	$b_k = \beta(k_\infty/\phi)^{-0.5}$
Civan	$b_k = 0.0094(k_\infty/\phi)^{-0.5}$

8.2.3 Slip Flow

In low-permeability formations (such as shale gas reservoirs) or when the pressure is very low, gas slip flow (or Klinkenberg effect) cannot be ignored when studying gas transport in tight reservoirs. Under such kind of flow conditions, gas permeability depends on gas pressure which can be expressed as follows:

$$\vec{\mathrm{k}_\mathrm{a}} = \vec{\mathrm{k}_\mathrm{i}}\left(1 + \frac{\mathrm{b}_k}{\mathrm{p}}\right) \tag{8.3}$$

where $\mathrm{k_a}$ is the apparent permeability, $\vec{\mathrm{k}_\mathrm{i}}$ is the intrinsic permeability, b is the slip coefficient. Many scholars have proposed different expressions for the slip coefficient which are listed in the Table 8.1.

8.2.4 Gas Adsorption and Desorption

Gas desorption occurs when pressure drop exists in organic grids (Kerogen). With free gas production, the pressure in the pores will decrease, which results in the pressure difference between the bulk matrix and the pores. Due to this pressure drop, gas will desorb from the surface of bulk matrix. The most commonly used empirical model describing sorption and desorption of gas in shale and providing a reasonable fit to most experimental data is the Langmuir single-layer isotherm model. Based on the Langmuir equation (Eq. (8.4)), the amount of gas adsorbed on the rock surface can be computed.

$$V_{ads} = \frac{V_L \cdot p}{p_L + p} \tag{8.4}$$

where V_L is the gas Langmuir volume denoting the amount of gas adsorbed at infinite pressure p_∞, scf/ton; P_L is the gas Langmuir pressure corresponding to the pressure at which half of the Langmuir volume $\mathrm{v_L}$ is reached, psi; P_g is the in-situ gas pressure in the pore system, psi. Langmuir volume means the maximum amount of gas that can be adsorbed on the rock surface under infinite pressure. Langmuir pressure is the pressure when the amount of gas adsorbed is half of the Langmuir volume. These two parameters play important roles in the gas desorption process. For different shale gas reservoirs, the differences of Langmuir volume and Langmuir pressure lead to distinct trend of gas content.

The adsorption gas volume per unit bulk volume can be expressed in Eq. (8.5):

$$q_a = \frac{\rho_s M_g}{V_{std}} q_{std} = \frac{\rho_s M_g}{V_{std}} \frac{V_L p_m}{p_L + p_m} \tag{8.5}$$

where q_a is the adsorption gas per unit area of shale surface, kg/m^3; V_{std} is the mole volume under standard condition (0 °C, 1 atm), m^3/mol; q_{std} is the adsorption volume per unit mass of shale, m^3/kg; V_L is the Langmuir volume, m^3/kg; p_L is the Langmuir pressure, Pa; ρ_s is the density of shale core, kg/m^3.

8.3 Apparent Gas Permeability in the Matrix and Fracture System

8.3.1 Apparent Gas Permeability in the Shale Matrix

Some empirical models have been developed to account for slip-flow and Knudsen diffusion in the form of apparent permeability, including correlations developed from the Darcy matrix permeability; correlations developed based on flow mechanisms; correlations as a function of Knudsen number based on Beskok and Karniadakis's work (Beskok and Karniadakis); and semi-empirical analytical models by [1]. This research adopted Javadpour's method which considered slip flow, viscous flow and Knudsen diffusion [15]. The apparent permeability is given as follows:

$$\vec{k}_m = \vec{k}_{mi}\left(1 + \frac{b_m}{p_m}\right) \tag{8.6}$$

$$b_m = \frac{16\mu}{3000r}\left(\frac{8\pi RT}{M}\right)^{0.5} + \left(\frac{8\pi RT}{M}\right)^{0.5}\frac{\mu}{r}\left(\frac{2}{\alpha} - 1\right) \tag{8.7}$$

where α is the tangential momentum accommodation coefficient which characterizes the slip effect. α is a function of wall surface smoothness, gas type, temperature and pressure which varies in a range from 0 to 1. For simplification, in this chapter, it is set as 0.8.

8.3.2 Apparent Gas Permeability in the Shale Fractures

In this study, Knudsen diffusion and viscous flow are considered for gas flow in the frature system. The mass flux can be expressed as the summation of the two mechanisms. Therefore,

$$J_f = J_{fv} + J_{fk} = -\frac{\rho_f k_{fi}}{\mu_g}(\nabla p_f) - \frac{\rho_f D_{kf}}{p_f}(\nabla p_f) \tag{8.8}$$

or

$$J_f = -\frac{k_{fi}\rho_f}{\mu_g}\left(1 + \frac{b_f}{p_f}\right)(\nabla p_f) \tag{8.9}$$

Using the form of conventional Darcy flow equation, the fracture apparent permeability can be expressed as:

$$\vec{\mathrm{k}}_{\mathrm{f}} = \vec{\mathrm{k}}_{\mathrm{fi}}\left(1 + \frac{b_f}{p_f}\right) \tag{8.10}$$

where

$$b_f = \frac{D_{kf}\mu_g}{k_{fi}} \tag{8.11}$$

$$D_{kf} = \frac{4k_{fi}}{2.8284\sqrt{\frac{k_{fi}}{\varphi_f}}}\sqrt{\frac{\pi RT}{2M_g}} \tag{8.12}$$

where p_f is the fracture pressure, k_{fi} is the initial fracture permeability, k_f is the apparent fracture permeability, b_f is the Klinkenberg coefficient for the fracture system, D_{kf} is the Knudsen diffusion coefficient for the fracture system, φ_f is the fracture porosity.

8.4 Numerical Simulation Model for Shale Gas Reservoirs with MsFHW

8.4.1 Model Assumption

Assumptions are as follows: (1) the flow is isothermal; (2) there are only one component, one phase flow in the shale reservoir; (3) the gas diffuse quickly and can achieve phase equilibrium instantaneously; (4) Formation rocks are incompressible and the porosity change due to rock deformation is ignored; (5) The gas in the shale reservoir can be assumed ideal gas with gas deviation factor equal to 1; (6) Gas adsorption–desorption kinetics obey Langmuir isotherm equation.

8.4.2 Mathematical Model

(1) Continuity equation

In the dual porosity model, there exist two mass balance equations corresponding for fracture and matrix systems respectively as indicated by Warren and Root [16]. The diagram of dual porosity model is shown in the Fig. 8.1. The continuity equation for the matrix and fracture system are as follows:

In the matrix:

$$\left(\frac{dM}{dt}\right)_m + \left(\nabla \cdot (\rho \vec{v})\right)_m = Q_m \tag{8.13}$$

In the fracture:

$$\left(\frac{dM}{dt}\right)_f + \left(\nabla \cdot (\rho \vec{v})\right)_f = Q_f \tag{8.14}$$

The first term on the left side M is the mass accumulation term; the second term on the left side $\vec{v}$ is the flow vector term; and the right side Q is the source/sink term.

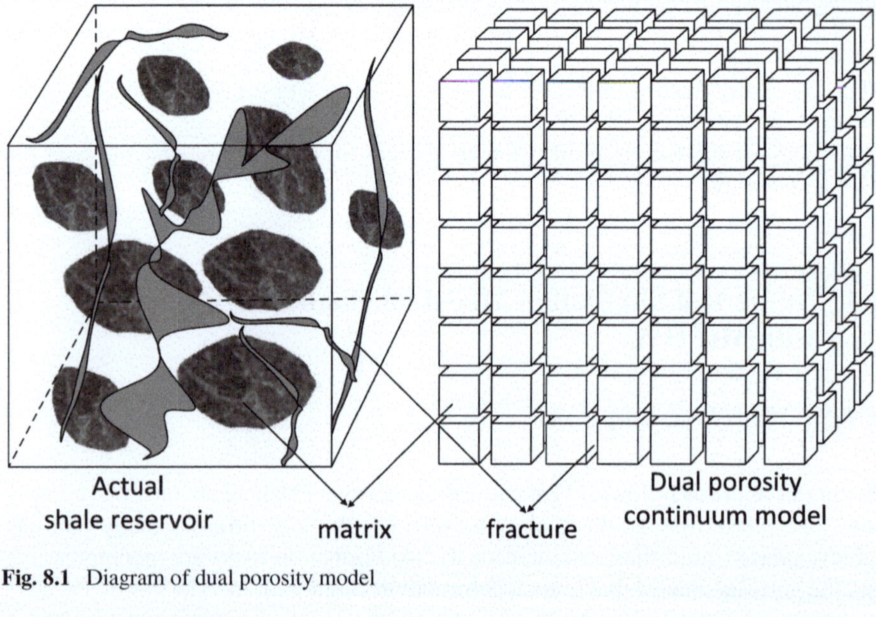

Fig. 8.1 Diagram of dual porosity model

In the matrix, there are adsorption gas and free gas co-existed. So, in the matrix, the mass accumulation term:

$$\mathrm{M} = \phi \sum_{\beta} \mathrm{S}_{\beta}\rho_{\beta} = \mathrm{M}_{\mathrm{free}} + M_{adsorp} = \phi \sum \mathrm{S}_{\mathrm{g}}\rho_{\mathrm{g}} + (1-\phi)\sum q_a \tag{8.15}$$

Based on the Equation of State (EOS) and Eq. (8.5), Eq. (8.15) can be transformed as follows:

$$\left(\frac{\mathrm{dM}}{\mathrm{dt}}\right)_{\mathrm{m}} = \frac{\partial(\rho_{\mathrm{g}}\phi)}{\partial \mathrm{t}} + \frac{\partial\left[\mathrm{q}_{\mathrm{g}}(1-\phi)\right]}{\partial \mathrm{t}} = \left[\gamma\varphi_{\mathrm{m}} + \frac{(1-\varphi_{\mathrm{m}})\mathrm{M}_{\mathrm{g}}\mathrm{p}_{\mathrm{L}}\mathrm{V}_{\mathrm{L}}\rho_{\mathrm{s}}}{\mathrm{V}_{\mathrm{std}}(\mathrm{p}_{\mathrm{L}}+\mathrm{p}_{\mathrm{m}})^{2}}\right]\frac{\partial \mathrm{p}_{\mathrm{m}}}{\partial \mathrm{t}} \tag{8.16}$$

where $\gamma = \frac{\mathrm{M}}{\mathrm{ZRT}}$. In the fracture system, only free gas existed, so the mass accumulation term can be described as follows:

$$\left(\frac{\mathrm{dM}}{\mathrm{dt}}\right)_{\mathrm{f}} = \frac{\partial(\rho_{\mathrm{g}}\phi))}{\partial \mathrm{t}} = [\gamma\varphi_{\mathrm{f}}]\frac{\partial \mathrm{p}_{\mathrm{f}}}{\partial \mathrm{t}} \tag{8.17}$$

(2) Motion equation

In the matrix and fractures, the permeability can be characterized using the modified Darcy's model. The motion equations are as followed:

$$\vec{v_m} = -\frac{K_{mg}}{\mu_g}\left[\nabla(p_{mg} - \rho_g g D)\right] \tag{8.18}$$

$$\vec{v_f} = -\frac{K_{fg}}{\mu_g}\left[\nabla(p_{fg} - \rho_g g D)\right] \tag{8.19}$$

In the matrix, $K_{mg} = k_{mi}(1 + b_m/p_m)$, where b_m is shown in Eq. (8.7).
In the fracture, $K_{fg} = \mathrm{k}_{\mathrm{mi}}(1 + \mathrm{b}_{\mathrm{f}}/p_f)$, where b_{f} is shown in Eq. (8.11).

(3) Gas production from Multi-stage hydraulic fracturing horizontal well

In the horizontal well with multi-stage transverse hydraulic fractures, the fractures are the main flow channels for the fluid flow from the shale micro-fractures into the horizontal wellbore. So, the gas production of horizontal wellbore is the sum of the gas production from each hydraulic fractures. For gas production from one single fracture, we assumed each fracture as a horizontal well along the y direction. Peaceman's model was used in this study Peaceman [17]. The following equation assume the horizontal well is along the x direction:

$$q_p = \frac{k_f\rho_f}{\mu_g}\frac{2\pi d_x}{\ln(r_e/r_w)+s}\left(\overline{p_f} - p_{wf}\right) \tag{8.20}$$

where p_{wf} is the bottom-hole pressure, and r_e is the effective radius and can be expressed as:

$$r_e = \begin{cases} 0.208\,\Delta y & \text{when } \Delta y = \Delta z,\, K_y = K_z \\ 0.28\dfrac{\left[(K_y/K_z)^{1/2}\Delta z^2 + (K_z/K_y)^{1/2}\Delta y^2\right]^{0.5}}{(K_y/K_z)^{1/4} + (K_z/K_y)^{1/4}} & \text{other} \end{cases} \tag{8.21}$$

(4) Numerical simulation model

By put the motion equations into continuity equation, the complete numerical simulation model was derived which is composed of flow equation, boundary conditions and initial condition:

Equation in the matrix system:

$$\left[\gamma\varphi_{\mathrm{m}} + \frac{(1-\varphi_{\mathrm{m}})\mathrm{M_g p_L V_L}\rho_{\mathrm{s}}}{\mathrm{V_{std}}(\mathrm{p_L}+\mathrm{p_m})^2}\right]\frac{\partial \mathrm{p_m}}{\partial \mathrm{t}} - \nabla\cdot\left[\gamma\left[\frac{\mathrm{k_{mi}}(\mathrm{p_m}+\mathrm{b_m})}{\mu_{\mathrm{g}}}(\nabla \mathrm{p_m})\right]\right] = -\mathrm{T} \tag{8.22}$$

Equation in the fracture system:

$$[\gamma\varphi_{\mathrm{f}}]\frac{\partial \mathrm{p_f}}{\partial \mathrm{t}} - \nabla\cdot\left[\gamma\left[\frac{\mathrm{p_f k_f}(\mathrm{p_f}+\mathrm{b_f})}{\mu_{\mathrm{g}}}(\nabla \mathrm{p_f})\right]\right] = \mathrm{T} - \mathrm{Q_p} \tag{8.23}$$

where T is the gas transfer between the matrix and fracture systems, which is represented in the Eq. (8.24):

$$\mathrm{T} = \frac{\mathrm{k_m}\rho_{\mathrm{g}}\sigma(\mathrm{p_m}-\mathrm{p_f})}{\mu} \tag{8.24}$$

where $\sigma = 4\left(\frac{1}{\mathrm{L_x^2}} + \frac{1}{\mathrm{L_y^2}} + \frac{1}{\mathrm{L_z^2}}\right)$ is the crossflow coefficient between the matrix and fracture systems, and $\mathrm{L_x}$, $\mathrm{L_y}$, $\mathrm{L_z}$ are the fracture spacing in the x, y, z directions. In the fracture system, there exists a source term which flows into fracture from the matrix and a sink term which flows out of fracture into the wellbore. Q_p is the sum of gas production from the fractures of them multi-stage hydraulic fractured horizontal well.

Initial condition: $\mathrm{p_m}|_{\mathrm{t}=0} = \mathrm{p_f}|_{\mathrm{t}=0} = \mathrm{p_i}$.

Boundary condition for matrix: $\mathrm{F_m}\cdot \mathrm{n}|_{\Gamma_1} = 0\left(\frac{\partial \mathrm{p}}{\partial \mathrm{n}}|_{\Gamma_1} = 0\right)$

Boundary condition for fracture: $\mathrm{F_f}\cdot \mathrm{n}|_{\Gamma_1} = 0\left(\frac{\partial \mathrm{p}}{\partial \mathrm{n}}|_{\Gamma_1} = 0\right)$, $\mathrm{p_f}(\mathrm{x},\mathrm{y},\mathrm{t})|_{\Gamma_2} = \mathrm{p_{wf}}$.

8.5 Sensitivity Parameters Analysis

In order to conduct sensitivity analysis, we constructed a shale gas reservoir model with a volume of 2500 ft × 2000 ft × 300 ft, which is called synthetic model. The parameters used in the synthetic model have been listed in the Table 8.2. In the middle of the reservoir, there is a horizontal well with six stages of hydraulic fractures which is shown in Fig. 8.2. Sensitivity analysis in this chapter was constructed based on the synthetic model and two categories of parameters have been studied on the production performance of shale gas reservoir: the reservoir parameters and the hydraulic fracture parameters. For the reservoir parameters, we have studied the effect of matrix permeability and matrix porosity on the production performance of the shale gas reservoir. For the hydraulic fracture parameters, two parameters have also been studied: fracture spacing and fracture half-length.

8.5.1 Effect of Matrix Permeability

Shale gas reservoirs are typical unconventional reservoirs with ultra-low permeability from 10 to 1000 nanodarcies (10^{-6} mD). According to the literature review, it can be found that matrix permeability of shale gas reservoirs in the U.S. are in the range of 10 nD (1 nD = 10^{-9} Darcy = 10^{-6} mD) to 1000 nD. The permeability distribution is shown in Fig. 8.3. Based on the reviewed range of matrix permeability, sensitivity

Table 8.2 Parameters used in the synthetic model for production performance study

Parameters	Barnett Shale case	Unit
Model dimensions (L × W × H)	2500 × 2000 × 300 (762 × 609.6 × 91.44)	ft (m)
Initial reservoir pressure	2400 (1.65 × 10^7)	Psi (Pa)
Bottom-hole pressure	500 (3.45 × 10^6)	Psi (Pa)
Production period	20 (6.31 × 10^8)	Year (s)
Reservoir temperature	200 (93.3)	°F (°C)
Gas viscosity	0.02 (0.00002)	cP (Pa s)
Bulk density	158 (2530.92)	lb/ft^3 (kg/m^3)
Reservoir top depth	6800 (2072.6)	ft (m)
Langmuir pressure	650 (4.48 × 10^6)	Psi (Pa)
Langmuir volume	100 (0.0028)	SCF/ton (m^3/kg)
Fracture conductivity	9 (1.22 × 10^{-14})	md ft (m^2 m)
Matrix permeability	0.00035 (6.125 × 10^{-19})	md (m^2)
Matrix porosity	0.06	Fraction
Horizontal wellbore length	1000 (304.8)	ft (m)

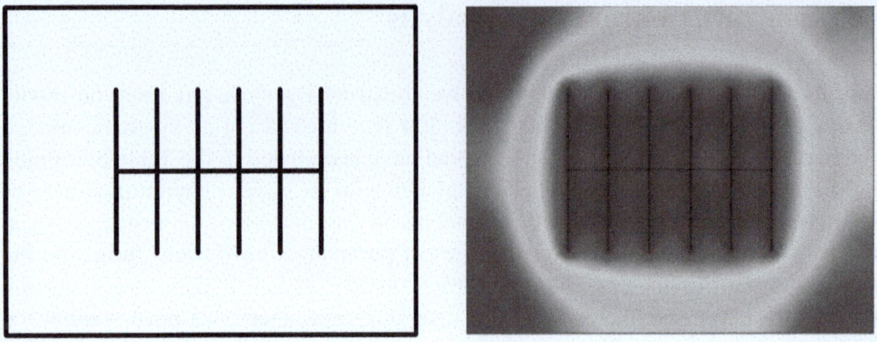

Fig. 8.2 Synthetic model for production performance study

analysis of matrix permeability on production performance of shale gas reservoirs has been conducted. In this study, we have considered five cases with matrix permeability close to the reviewed matrix permeability range. The matrix permeability of the five cases are: 10, 200, 500, 800, and 1000 nD.

The effect of matrix permeability on cumulative gas production and gas production rate are shown in Fig. 8.4. From this figure, it can be found that the effect of matrix permeability on shale gas production performance is not significant. Also, it can be found that in the early period, the matrix permeability barely has influence on gas production. In the late period of gas production, the difference began to appear. However, the difference is not significant. When the matrix permeability changes from 10 to 1000 nD, the cumulative gas production in 20 years only increases from 2540 to 2750 MMSCF, which is about 7.65% increase.

Fig. 8.3 Frequency distribution diagram of shale matrix permeability

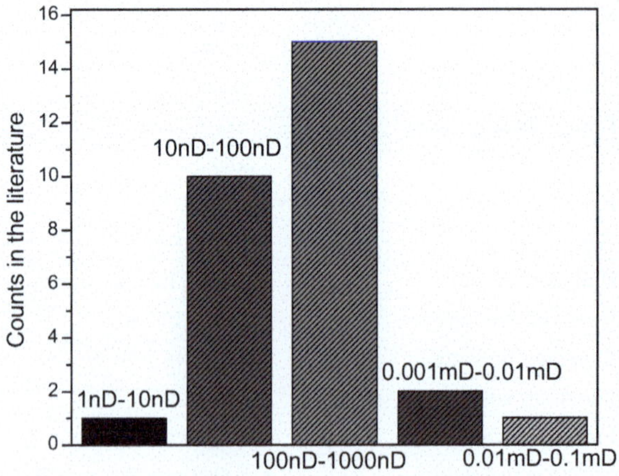

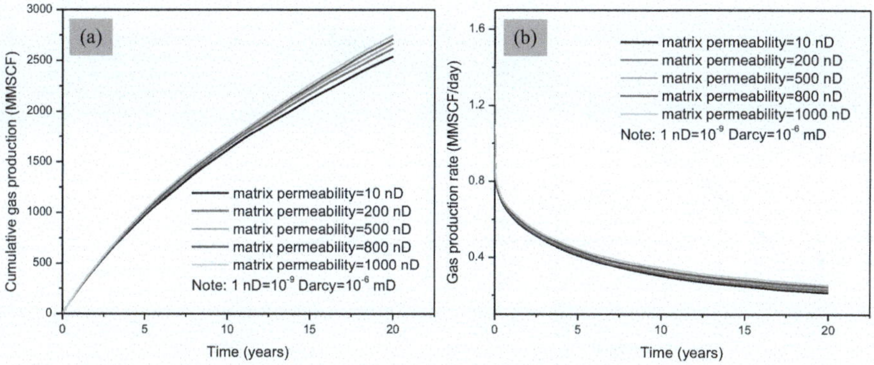

Fig. 8.4 Effect of matrix permeability on the production performance of shale gas reservoirs. **a** Cumulative gas production versus time; **b** gas production rate versus time

8.5.2 Effect of Matrix Porosity

Sensitivity analysis of matrix porosity on production performance of shale gas reservoirs has also been conducted based on the reviewed range of matrix porosity, which is shown in Fig. 8.5. It can be found that matrix porosity are in the range of 2–8%. So, in this study, we have considered five cases with matrix porosity close to this range. The matrix porosity of the five cases are: 2, 4, 6, 8, and 10%. The effect of matrix porosity on cumulative gas production and gas production rate are shown in Fig. 8.6.

From Fig. 8.6, it can be found that the effect of matrix porosity on the production performance of shale gas reservoirs is also not very significant. When the matrix porosity changes from 2 to 10%, there is no significant change in the early and late

Fig. 8.5 Frequency distribution diagram of shale matrix porosity

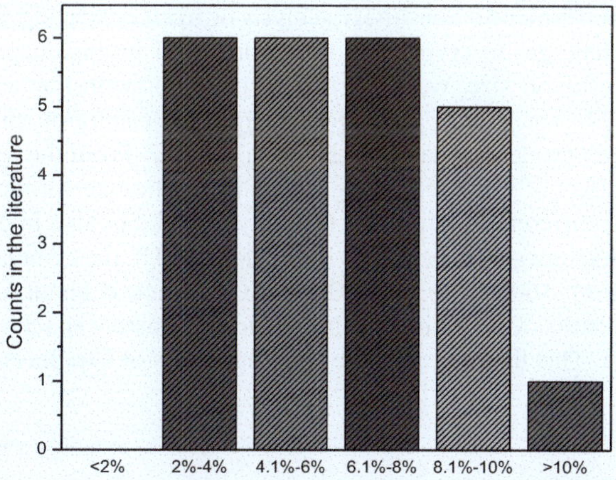

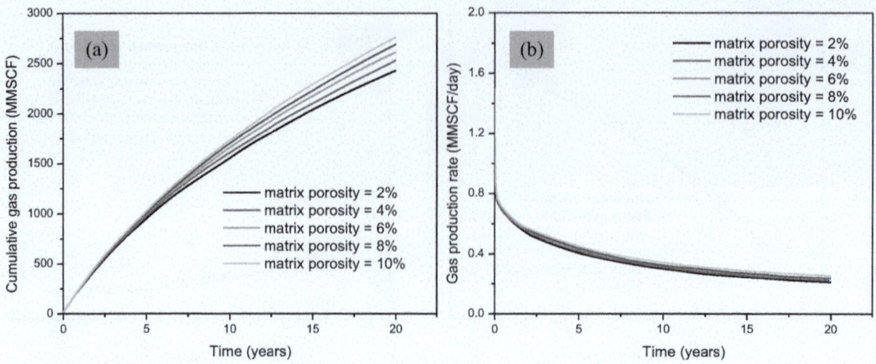

Fig. 8.6 Effect of matrix porosity on the production performance of shale gas reservoirs. **a** cumulative gas production versus time; **b** gas production rate versus time

production period. The cumulative gas production changes from 2431.49 to 2759.48 MMSCF, which is 12% increase.

8.5.3 *Effect of Fracture Spacing*

When there are multiple transverse fractures in shale gas reservoirs, it is important to consider the effect of fracture spacing. Cipolla et al. [18] found that in order to achieve the commercial production rates and optimum depletion of shale gas reservoir, the fracture spacing should be minimized. However, when the fracture spacing is too small, the interaction between fractures will become more apparent, which is not good for production. Also, more fractures will require more investment to create fractures. So, it is important to study the effect of fracture spacing on shale gas production performance. Sensitivity analysis of fracture spacing on production performance of shale gas reservoirs has been conducted in this chapter. Due to limitations of the reservoir size, we only studied four cases with fracture spacing equal to 100, 200, 300, and 400 ft. The effect of fracture spacing on cumulative gas production and gas production rate are shown in Fig. 8.7. From the figure, it can be found that the effect of fracture spacing on gas production performance is very significant. There is difference not only in the early period, but also in the late period. And when the fracture spacing changes from 100 to 400 ft, the cumulative gas production increases from 1606.47 to 3342.32 MMSCF, which is about 52% increase. Also, from the results, we can find that the smaller the fracture spacing, the more the gas production is. This findings coincide with the results of Cipolla et al. [18].

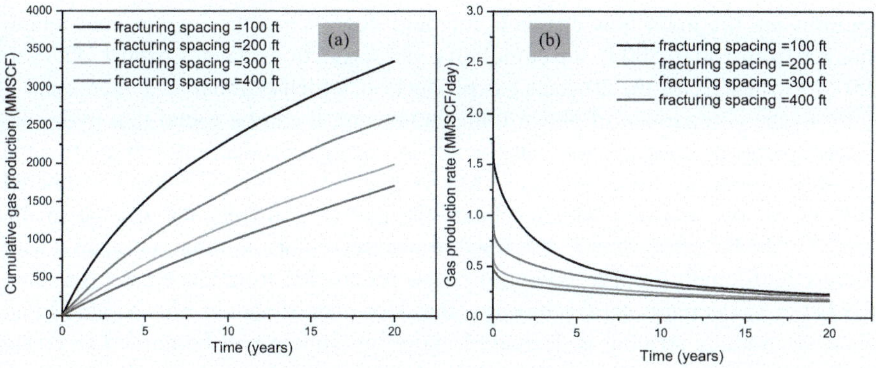

Fig. 8.7 Effect of fracture spacing on the production performance of shale gas reservoirs. **a** cumulative gas production versus time; **b** gas production rate versus time

8.5.4 *Effect of Fracture Half-Length*

Fracture half-length is the horizontal distance from horizontal wellbore to the end of hydraulic fracture. Fracture half-length is important in deciding production from multi-stage hydraulic fracturing horizontal well as this is the main high conductivity channels for fluid flow from the reservoir into the horizontal wellbore. Sensitivity analysis of fracture half-length on production performance of shale gas reservoirs has been conducted in this study. We have considered five possible cases with considering the limitation of the reservoir size. The fracture half-length of the five cases are: 100, 200, 300, 400, and 500 ft. The effect of fracture half-length on cumulative gas production and gas production rate are shown in Fig. 8.8.

From Fig. 8.8, it can be found that the larger the fracture half-length, the higher the cumulative gas production and the gas production rate. This is because the hydraulic

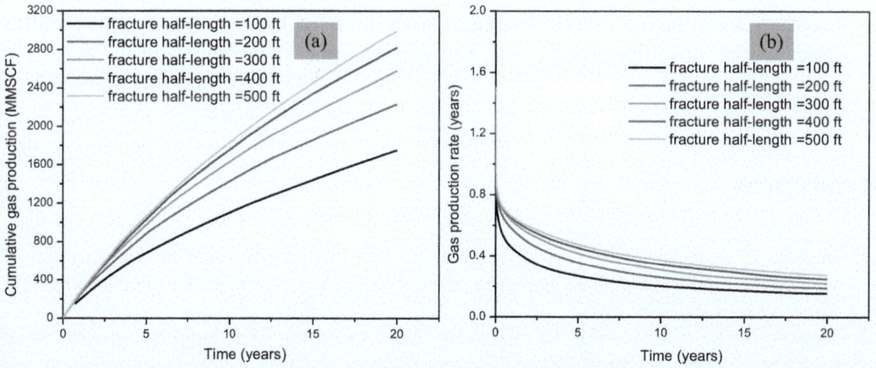

Fig. 8.8 Effect of fracture half-length on the production performance of shale gas reservoirs. **a** cumulative gas production versus time; **b** gas production rate versus time

fractures are the main flow channels for fluid flow. When the fracture half-length changes from 100 to 500 ft, the cumulative gas production changes from 1747.79 to 2991.15 MMSCF for the 20 years gas production, which is about 42% increase.

From the comparison of above four parameters, it can be found that hydraulic fracture parameters are the dominate factors in shale gas production with multi-stage hydraulic fracturing horizontal well. Matrix porosity is more dominant compared with matrix permeability which denotes the matrix is mainly the storage media instead of the fluid flow media. And fracture spacing is more dominate compared with fracture half-length. In order to optimize gas production from shale gas reservoirs, it is better to decrease the fracture spacing. However, it should also be noted that when the fracture spacing is decreased, more investment will be needed to create the fractures. So, the optimal plan will be the balance between technology and the economic investment.

8.6 Remark

In this chapter, we make an attempt to study the influence of various parameters on the production performance of ultra-low permeability shale gas reservoirs with multi-stage hydraulic fracturing horizontal well.

(1) A dual porosity model was constructed to simulation gas production from shale reservoirs with multi-stage hydraulic fracturing horizontal well with considering multiple flow mechanisms, including: viscous flow, slip flow, Knudsen diffusion, and gas desorption.
(2) A synthetic model has been constructed to analyze the effect of reservoir parameters and hydraulic fracture parameters on gas production performance based on the reviewed ranges of corresponding parameters.
(3) According to the sensitivity analysis, it can be found that hydraulic fracture parameters are the dominate factors which affect the gas production. Fracture spacing is more dominant compared with the fracture half-length. And matrix porosity is more dominant compared with the matrix permeability.

References

1. Moridis, G. J., Blasingame, T. A., & Freeman, C. M. (2010). Analysis of mechanisms of flow in fractured tight-gas and shale-gas reservoirs. In *SPE Latin America and Caribbean Petroleum Engineering Conference*. SPE.
2. Watson, A. T., et al. (1989). An analytical model for history matching naturally fractured reservoir production data. In *SPE Oklahoma City Oil and Gas Symposium/Production and Operations Symposium*. SPE.
3. Bustin, R. M., et al. (2008). Impact of shale properties on pore structure and storage characteristics. In *SPE Shale Gas Production Conference*. SPE.

4. Ozkan, E., Raghavan, R., & Apaydin, O. G. (2010). Modeling of fluid transfer from shale matrix to fracture network. In: *SPE Annual Technical Conference and Exhibition?* SPE.
5. Connell, L. D., & Lu, M. (2007). A dual-porosity model for gas reservoir flow incorporating adsorption behaviour—Part II. Numerical algorithm and example applications. *Transport in Porous Media, 69*(2), 139–158.
6. Renyi, C., Linsong, C., & Peiqing, L. (2017). Mathematical model of shale gas reservoirs coupled by desorption, stress sensitivity and fractures finite conductivity. *Physical and Numerical Simulation of Geotechnical Engineering, 26*, 27.
7. Pruess, K. (1992). *Brief guide to the MINC-method for modeling flow and transport in fractured media.*
8. Wu, Y.-S., et al. (2009). A multi-continuum model for gas production in tight fractured reservoirs. In *SPE Hydraulic Fracturing Technology Conference and Exhibition.* SPE.
9. Civan, F., Rai, C. S., & Sondergeld, C. H. (2011). Shale-gas permeability and diffusivity inferred by improved formulation of relevant retention and transport mechanisms. *Transport in Porous Media, 86*, 925–944.
10. Dehghanpour, H., & Shirdel, M. (2011). A triple porosity model for shale gas reservoirs. In *SPE Canada Unconventional Resources Conference?* SPE.
11. Yan, B., Wang, Y., & Killough, J. E. (2016). Beyond dual-porosity modeling for the simulation of complex flow mechanisms in shale reservoirs. *Computational Geosciences, 20*, 69–91.
12. Mayerhofer, M. J., et al. (2006). Integration of microseismic fracture mapping results with numerical fracture network production modeling in the Barnett shale. In *SPE Annual Technical Conference and Exhibition?* SPE.
13. Gong, B., et al. (2011). Detailed modeling of the complex fracture network of shale gas reservoirs. In *SPE Middle East Unconventional Resources Conference and Exhibition.* SPE.
14. Kast, W., & Hohenthanner, C.-R. (2000). Mass transfer within the gas-phase of porous media. *International Journal of Heat and Mass Transfer, 43*(5), 807–823.
15. Javadpour, F. (2009). Nanopores and apparent permeability of gas flow in mudrocks (shales and siltstone). *Journal of Canadian Petroleum Technology, 48*(08), 16–21.
16. Warren, J. E., & Root, P. J. (1963). The behavior of naturally fractured reservoirs. *Society of Petroleum Engineers Journal, 3*(03), 245–255.
17. Peaceman, D. W. (1983). Interpretation of well-block pressures in numerical reservoir simulation with nonsquare grid blocks and anisotropic permeability. *Society of Petroleum Engineers Journal, 23*(03), 531–543.
18. Cipolla, C. L., et al. (2010). Reservoir modeling in shale-gas reservoirs. *SPE Reservoir Evaluation & Engineering, 13*(04), 638–653.

Chapter 9
Conclusions

This book has provided a comprehensive exploration of shale gas production, covering foundational concepts, gas transport dynamics, and hydraulic fracturing techniques in multi-stage horizontal wells. Our goal has been to equip readers with both theoretical knowledge and practical insights, making this resource beneficial for students, researchers, and industry professionals. Although we have aimed to cover the essential aspects of shale gas engineering, time constraints have limited the depth of our analysis in certain areas. We kindly ask readers to consider this, and we appreciate your understanding of any limitations or areas where further elaboration could have been beneficial.

Looking forward, several areas stand out for potential improvement and future research. For instance, refining gas transport models to incorporate more complex multi-phase and multi-component interactions within diverse pore structures would enhance our understanding of reservoir behavior. While this text has explored two-component and two-phase transport models, we believe future studies could expand upon these models by accounting for additional variables such as pore shape variations and dynamic pressure conditions.

Another promising direction lies in the continued development of hydraulic fracturing techniques to balance extraction efficiency with environmental considerations. Real-time monitoring, adaptive fracturing, and data-driven methods like machine learning could play a significant role in optimizing production strategies, reducing environmental impact, and improving predictive modeling capabilities.

Despite the time limitations, this work may have been somewhat rushed, and we kindly ask for your understanding if certain sections lack detail or clarity. We hope this book serves as a valuable resource and sparks further exploration and innovation in shale gas production. The field offers immense potential, and through ongoing research, collaboration, and technological advancements, we can address the challenges and make meaningful strides toward sustainable and efficient shale gas extraction.

C. Guo and Z. Yang, *Shale Gas Production: Concept, Models, and Techniques*, Petroleum Engineering, https://doi.org/10.1007/978-3-031-84552-9_9

Acknowledgements

The completion of this book would not have been possible without the support and encouragement of numerous individuals. We extend our heartfelt thanks to our colleagues and students in China University of Geosciences (Wuhan), who have engaged with us in discussions and shared their insights over the years. Your contributions have been invaluable in shaping the content of this text.

We are especially grateful to our research teams for their dedication and perseverance in exploring the complexities of shale gas production. A special thanks to Prof. Baojun Bai and Prof. Mingzhen Wei from Missouri University of Science and Technology for their guidance and expertise, which greatly influenced the development of our ideas and models.

We would also like to acknowledge the assistance of our editorial team, including Anu Mary Peter; Anthony Doyle; Amudha Vijayarangan, et al. They have provided critical feedback and support throughout the writing process. Their expertise in refining our manuscript has been instrumental in bringing this project to fruition.

Finally, we thank our families for their unwavering support and patience during the long hours dedicated to this endeavor. Your love and encouragement have been our greatest motivation.

C. Guo and Z. Yang, *Shale Gas Production: Concept, Models, and Techniques*, Petroleum Engineering, https://doi.org/10.1007/978-3-031-84552-9

The manufacturer's authorised representative in the EU is Springer Nature Customer Service Centre GmbH, Europaplatz 3, 69115 Heidelberg, Germany. If you have any concerns regarding our products, please contact ProductSafety@springernature.com

Printed and bound by CPI Group (UK) Ltd, Croydon, CR0 4YY
01/10/2026
02225987-0008

TABLE OF CONTENTS

INTRODUCTION

The inspiration for this book began in 1973 at age 10 in the basement of my home in Wilmette, Illinois. It was here I became "master of the universe," at least on a HO model train scale. It all began with a standard Christmas gift of a train set which soon expanded into a 25x20 foot layout. Cold Chicago winters were the impetus for making a scenic panorama where the trains ran on time.

My grandfather, Thilo Max Best, often visited on birthdays and holidays. One such time he brought along a vast collection of photographs from the Jacob Best Coal Company of Chicago. The pictures were entrusted to me, and I put them to good use as the inspiration for a Best Coal Company in miniature as part of my train layout. First I did the Mayfair Yard and then the Edgewater Yard. Pa, as I called my grandfather, served as my technical advisor. The layout grew ever more elaborate with the addition of the Appalachian Mountains.

I had a partner in all this, my younger brother Robert. Eventually, we agreed on a leveraged buyout. Some money changed hands, along with a baseball hit by the Cubs' Jose Cardenal. I was then the sole proprietor of a world where coal still ruled as king.

In 1974 we took a family trip to Maryland and visited the Pennsylvania Glass Sand Corporation, once owned by the Bridges' side of the family. The tour of the Corporation fascinated me to the point of collecting samples of raw silica and photographing the yellow thirty-ton trucks used to move rock. I was an eleven-year old with a plan. Back home in Wilmette, I added a silica mining pit to the layout. My ancestral past in rock and coal now showed in HO.

After my grandparents died, the railroad layout reminded me how their lives had been shaped by hard work and an entrepreneurial spirit. My subsequent education and entry into the Presbyterian Ministry did not lessen this fascination with my family history. In 1997 I began writing two family histories from the time my ancestors arrived in America. It took me five years to write the two books that chronicle the story of the Bests and the Bridges.

What follows is the story of the Bests-and their business-from 1869 to 1964. The Best Coal Company would not be confused with Ford or US Steel, although it provided an important product for Chicago, both in terms of sales and or employment. But drawing on a faith both secular and religious, the Bests stand out as an example of Americans able to create a fortune.

The dynamic relationship between economic and religious life has influenced Americans' political and social well being since the arrival of the Mayflower in 1620. The Best's were deeply ethical people, and their businesses were affected by their ethics and religious beliefs. At the foundation of the American Dream there has always been an ideal. Intimately associated with it are faith, the accumulation of wealth and individual freedom.

This story also examines the question of how the entrepreneurial spirit inspired these individuals, from immigrant to third generation, to succeed. Something drove them on regardless of circumstances. It was courage and the times, along with those personalities that so marked the Bests.

This narrative begins with an immigrant from Prussia who made his way to America in the nineteenth century. Many of these German immigrants worked their way west from the ports of Baltimore and New York and north from New Orleans. In 1869 Jacob Best from the village of Hettenleidelheim joined them. He settled in Chicago and worked as a saloonkeeper.

Jacob Best, the son of this saloonkeeper, became a leader in Chicago's retail/wholesale coal business. Starting in 1909, his company provided the energy to heat homes, apartments and businesses as well as to power the factories which generated so much of the city's wealth. As a Chicago philanthropist, my great grandfather sought to improve the lives of his customers and neighbors alike.

In 1960 the Best and the Bridges families were united through the marriage of my mother, Barbara Ann Best, granddaughter of Jacob Best, and my father, Shelby Powell Bridges, son of Henry P. Bridges. In a way, this blending of families, of adding one story to another, is what I was trying to work out in the basement. I think this book begins to capture the wonder better than a train layout could.

ACKNOWLEDGEMENTS

I could not have written my two family history books over the course of the past five years without the assistance of my professors at the University of Chicago and its Divinity School. I would especially like to thank the following people for their part in teaching me how to research and write American religious history: Dr. Martin Marty, Dr. Clark Gilpin and Dr. Catherine Brekus. I am also grateful for the encouragement and advice of my good friend Dr. John Mulder, president of my alma mater, Louisville Presbyterian Theological Seminary.

A family history would be impossible without family support, and in that I have been fortunate. My father Powell Bridges and my cousin, Jerry Best, have been instrumental in helping me to find primary documents and research materials pertinent to the manuscript. My mother, Barbara Best Bridges, has listened patiently and helped to guide the writing of this, the Best story.

Mr. Richard Gaynor enabled me to move forward thanks to his exceptional computer skills. Richard made possible the first, desktop published format. Other people have graciously listened to me and encouraged the project. Among these is Linda DeLeon, at The University Club of Chicago, a fellow author. She has listened and together we have discussed the plight of authorship. In addition Cindy Wietell at ABS Service did an excellent job of typing all the primary research documents needed to complete the project.

Mr. Toby Tobias, my second cousin, became an invaluable content editor. His love for genealogical research proved to be "just what the doctor ordered" at a time when my writing stamina needed his support. Dr. Douglas Bukowski, Ph.D. in American History from the University of Illinois at Chicago worked to edit this book. Dr. Ricia Chansky, Ph.D. from The George Washington University also did an excellent job as a Copy Editor. Lastly, thanks to all those readers who freely gave valuable insight into the writing and rewriting of this book!

CHAPTER 1

JACOB BEST FROM PRUSSIA TO CHICAGO'S SOUTH SIDE TO POINTS NORTH, 1835-1887:

A German Saloon Entrepreneur in Chicago

OVER HERE

On August 19, 1869, the marriage book for Hettenleidelheim Kirchenbuch recorded that, "Jacob Best and his bride Katherine Ness Best walked out of the Catholic Church in the small town of Hettenleidelheim, Prussia" as man and wife. Jacob had just turned thirty-four and Katherine was all of twenty-five. The groom had grown up in the neighboring town of Ramsen. Jacob was a short man in stature, with a rounded face and hair neatly parted to the left. He was well dressed, and his bow ties gave him a proper look for the time and place. Katherine mirrored his round-faced countenance, and she had long arms and fingers. She possessed a certain simple elegance. They made a fitting couple.

But Prussia did not allow for a very happy home. The kingdom had been at war on and off for nearly 150 years, and the Age of Napoleon did little to bring peace. After Waterloo, only four cities-Bremen, Frankfurt, Hamburg, and Lubeck-enjoyed any semblance of self-government. Small agrarian villages like Ramsen and Hettenleidelheim remained under feudal rule. By 1830 the Prussian people had grown dissatisfied with their living conditions and their economic well being, or lack thereof.

The forces of modernization and industrialization only increased this general sense of anxiety, both in Prussia and the other still-independent German states. Then in 1862, Otto von Bismarck was selected as Prime Minister of Prussia. Bismarck wanted a united Germany, and he was willing to use the military to achieve it. Wars against Denmark in 1864 and Austria two years later furthered his ends, as did the establishment of the North German Confederation in 1867. The unification of Germany was nearing completion.

It would take the Franco-Prussian War of 1870-71 with the French surrender of Alsace and part of Lorraine for Bismarck to finish. Wilhelm I of Prussia was crowned Kaiser, and Imperial Germany under the leadership of Chancellor Bismarck became a world power. But this was not Jacob Best's future.

Jacob and Katherine had known each other since childhood. Shortly after their wedding day they boarded a steamship for America. According to government records, they arrived November 2, 1869. Had he stayed in Prussia, Jacob likely would have joined in the fight against France a year later. Such was the fate of many Prussians his age, but Jacob had different plans for himself. America was the land of opportunity the newlyweds wanted to experience. The bustling city of Chicago, with its large German population, presented itself as an obvious choice as their destination.

JACOB BEST 'JUST OFF THE BOAT'

Jacob and Katherine most likely arrived in Chicago by train, which was only fitting, since they were about to live in the railroad capital of the United States, if not the world. Chicago had been founded in 1833 and incorporated as a city four years later. The 1850 population of 30,000 jumped to over a million within a span of forty years. This extraordinary growth bore a distinctly German flavor. In 1900 some 440,000 residents of German descent lived in Chicago. "People talked of Chicago as a German city, rather than an Irish one," urban historian Perry Duis has noted.

According to one study, in Chicago "The experience of both Germans and other immigrants on their arrival at Chicago's train stations remained substantially the same over the decades: the immigrants were met either by relatives and friends or by agents of the railroads, the police, or the German Society, an immigrant-aid organization. There were also legions of runners trying to relieve the immigrants of their money by leading them to local restaurants and boardinghouses where they would be overcharged." Jacob and Katherine found the latter if not the former soon after they stepped off their train.

The *Arbeiter-Zeitung,* a German-language paper, accused both the German Society and the railroads of failing to deal with the problems newcomers faced. The paper was particularly critical of runners in the employ of some of the city's German innkeepers.

These men implied a connection to the society, won the trust of many immigrants, and proceeded to take their money. Victims soon learned that not all the risks associated with immigration ended with the voyage.

Jacob Best and wife Katherine Ness Best. If immigrants could afford one luxury, it was a picture portrait from the neighborhood photography studio, like this one on Archer Avenue.

CHICAGO: A NEW HOME FAR FROM PRUSSIA

But Jacob and Katherine survived the gauntlet of runners at the train station. The 1880 Census picks them up eleven years later living in a house at 2701 Cottage Grove Avenue. The couple was living in a German-Catholic neighborhood just south of downtown. The Census reports that Jacob worked as a bartender and was the father of four children: Tillie, born in 1870; George, 1872; Jacob 1878; and Maryanne, 1880.

The Germans who settled in Chicago were primarily artisans, shoemakers, tailors, blacksmiths, butchers, brewers, skilled workers and small businessmen. By 1880 skilled workers in Chicago made up 37.4 percent of all employed Germans. Many found work with the largest industry in the city, meat packers with such firms as Swift, Armour and Oscar Mayer. Another major employer in the early 1880s was the McCormick Harvester Works, home of the "reaper" that revolutionized agriculture. More than half the workers at the plant had German names.

Both native-born Chicagoans and newcomers saw their home as a kind of miracle-city that had sprung up on the shore of Lake Michigan almost overnight. As the poet Carl Sandburg understood, Chicago was the place where people brawled, played and worked with flair. In addition to being a rail center, Chicago in the nineteenth century served as one of the busiest ports in the world. Jacob Best, who lived only a few short blocks from the lakefront, could see the frenzy of commerce all around him. Some of that prosperity was about to come his way. After fourteen years of bartending on the South Side of Chicago, Best accumulated enough cash to buy his own establishment. In 1884 Jacob purchased the Gross Park Hotel from developer Samuel Gross. The hotel and surrounding community were located in what was then the suburb of Lake View. With a little bit of effort, Jacob had himself a base for business.

The secret to real estate is, of course, location. Jacob did well to purchase land in what was then north-suburban Lake View, which was annexed to Chicago in 1889. For the next seventy-five years, Lake View would be the heart of German Chicago. With luck and effort, Jacob could expect to do well by his countrymen with a business on the corner of Roscoe and Lincoln streets, in the growing area of North Chicago.

THE GROSS PARK HOTEL: BEST AS BUSINESSMAN

So began a family adventure on June 21, 1884: "This Indenture Witnesseth, that the Grantors Samuel E. Gross and Emily Gross his wife of the City of Chicago in the County of Cook and State of Illinois for and in consideration of the sum of Seven hundred and twenty five ($725) dollars in hand paid Convey and Warrant Jacob Best of the City of Chicago County of Cook and State of Illinois the following described real estate to wit: Lots nine (9) and twelve (12) in block three (3) in Gross Park addition to Chicago Subject to all taxes and assessments after the year 1880." On June 26, 1885, Jacob bought "Lots Twenty five (25) and Twenty six (26) in Block Six (6) in L. Turner's Subdivision for $1,400 dollars." These lots were across the street from the Gross Park Hotel in Lake View. Best now had his own saloon.

The Gross Park Hotel was designed in the style of a boarding-house/saloon from the old country. Best built his business on the traditional hospitality, German-style. In a city like Chicago, different ethnic groups were unlikely to share a drink or even enjoy the same kind of drink. To outsiders, "The variety of ethnic saloons seems confusing," writes historian Melvin Holli, "Every group had its favorite beverage, spoke its own language freely, told its own jokes over the bar, and enjoyed its own delicacies at the free lunch sideboard." Jacob set about providing for one (very large) ethnic group.

Best also ran one of Chicago's most controversial businesses, a saloon. Journalist and muckraker Lincoln Steffens offered a none-toosubtle joke about the men who operate such places, "The fastest way to empty the City Council chambers was to stand at the rear and shout, 'Your saloon's on fire!'" And yet the saloon represented a step toward the American Dream. Ownership provided financial independence and an important position in many immigrant communities. Unfortunately, the dream did not come risk-free. As a businessman, the saloonkeeper had to worry about such fixed costs as license fees, taxes and rent, not to mention liquor. And no matter how reputable the establishment, a saloonkeeper could not expect to buck a recession, or a competitor opening up down the street. Lenders did not care about the way high unemployment affected business. They just wanted their money.

The liquor business in Chicago dates to the city's beginning. William Haas and Andrew Sulzer, a pair of German New Yorkers, settled in Chicago and soon began selling their ale. After a decade of brewing they sold out to William Lill and William B. Ogden. Ogden, who became the first mayor of Chicago, sold his interests to milkman Michael Diversey, and the company was renamed Lill and Diversey. By 1854 there were at least six other ale plants, though Lill and Diversey were still the top producer. Valentin Busch and Michael Brand were two other Germans who started up an ale plant in Blue Island, south of Chicago. The 1859 Chicago Directory advertised their product as "A Large Stock of Extra Lager Beer always on hand." In time, such famous names as Pabst, Blatz and Schlitz would come to Chicago. However, Jacob was a faithful patron of the Busch & Brand Brewing Company.

Brewers.

BRISACH & HESSIMER, (Joseph Brisach and Louis Hessimer,) OAK bet GREEN BAY AND THE LAKE.

BRISACH & HESSIMER,
BREWERS OF
LAGER BEER,
OAK STREET,
BETWEEN GREEN BAY STREET AND THE LAKE.

BUSCH & BRAND, (Valentine Busch and Michael Brand,) CEDAR n THE LAKE.

BUSCH & BRAND,
LAGER BEER BREWERS,
CHICAGO AND BLUE ISLAND.
A large Stock of Extra Lager Beer always on hand.

BUCHER & HILLER, (John Bucher and George Hiller,) GREEN BAY opp CEMETERY.

BUCHER & HILLER,
BREWERS OF
SUPERIOR CHICAGO LAGER BEER,
AND DEALERS IN
Malt, Hops & everything connected with the Business.
GREEN BAY ROAD, OPPOSITE THE CEMETERY, CHICAGO.

HUCK JOHN A., WOLCOTT cor N. DIVISION.

Busch & Brand Brewing Company advertisement.
Beer was a business both big and controversial in Chicago.

Ironically, saloons were the first business in Chicago to rebound from the Great Fire of 1871. Living south of downtown, Jacob and Katherine escaped the ravages of the blaze, which spread north and west of Mrs. O'Leary's barn. Joseph Chesterfield Mackin, a barkeep affectionately known as Oyster Joe, was especially quick to reestablish his business. Mackin's saloon served as the unofficial headquarters of the Cook County Democratic Party, and he was alleged to be heavily involved in area prostitution and gambling. Perhaps of greater interest to Jacob Best, Mackin may have been the saloonkeeper responsible for that innovation known as the free lunch.

Oyster Joe was an entrepreneur of considerable insight. He constructed a comfortable library in his saloon to draw away customers from his rivals. This backroom was full of magazines and papers, the kind of material to attract gamblers or horse-betters. But such customers were not completely loyal to his saloon, so Mackin went a step further and offered a free hot oyster with every drink consumed in his saloon. Thus was born the free lunch, with Chicago saloons giving away as much as $26 million in food a year. Although it was not his intention, Mackin popularized a giveaway that had the end result of feeding the hungry.

Not only did Jacob Best have to follow Oyster Joe's lead and feed the customers, he had to deal with the so-called "blind pigs," illegal saloons that sought to avoid the $250 license fee. Bottled beer was yet another major challenge. A group of independent bottlers discovered that they could buy the beer in barrels and resell it in bottles from wagons that traveled the city. Beer wagons soon were making more money than the saloonkeepers. Critics complained how "the wagons go around and fill people up with beer right on the street. Some of them sell whiskey and cigars and are practically saloons on wheels. They pay no license and still compete with the saloonkeeper who has to pay $1,000 a year." With this innovation the brewers and the bottlers sold more beer, but The German Tavern Keepers Association, known as Alter Wirths Verein, protested and even recommended a boycott of Joseph Schlitz's beer sold from the wagons. The saloonkeepers soon adjusted to the competition by selling bottles themselves.

What attracted Jacob to such a challenging business? He thought it was a good way to make both a profitable and honorable living. In Chicago, newcomers depended on the saloon for far more than drink. The saloon became a vital communications center, and it was the place where newcomers could learn the ropes of daily life in the city. The bartender would direct people to the local aid societies, lawyers and jobs. He would even post listings of missing children. In a time before social workers there was the man behind the bar.

Because he was literate, Jacob could read the German or English letters his patrons brought to him; this skill brought him honor as well as business. His saloon also served as an unofficial post office while the basement became a storage area for newcomers' luggage and trunks; people arrived from the old country with his saloon address as the only one they knew in the city. Best opened the Gross Park Hotel saloon between 5 and 6 AM in order to serve the working man who might stop by for a morning bracer before heading off to work. The noontime crowd was equally important. They came in droves for a free lunch-and a not-so-free beer. During the afternoon hours, women sent their children into the bar for a bottle of beer or a little whiskey or wine as an ingredient in that evening's meal. By closing time, the barroom welcomed any and all laborers shuffling home after a long day at work.

The saloonkeeper was also a philanthropist, and the Chicago German Aid Society relied on Best for generous donations. It was good business to have the name of the Gross Park Hotel listed as a donor. Even the settlement house workers had to concede that the largess of saloonkeepers helped those who could not provide for themselves. In the case of a family emergency, saloonkeepers would pass a hat or leave a cigar box on the bar for patrons to fill. When Chicago police officer John Heubner died in 1882, a neighborhood bar at Milwaukee and Division streets spearheaded relief efforts, to the point that the mortgage on Heubner's house was retired. Jacob Best's saloon reflected his personality, and all the fixtures in the barroom were meticulously placed around the bar according to his taste. The breweries sold or rented fixtures (lamps, stained glass windows, beer taps) to thousands of saloons in exchange for exclusive agreements. More than a few of Best's competitors took advantage of breweries offering a monthly tab; payment was not always forthcoming. The so-called "good old days" included some sharp practices at all levels of the liquor business.

In 1885 the ratio of citizens per saloon in Chicago stood at 203 to 1; there were more bars in Chicago than in all fifteen states below the Mason Dixon Line. For Jacob Best those numbers translated into status both in the community and the profession. The Constitution and By-Laws of*The Saloon-Keepers' Union of Lake View* reveals a great deal about the business aspect.

The Constitution, Article I. states, "This Society shall be called the *Saloon-Keepers' Union* of the Town of Lake View, and its object shall be to protect and defend their common interests by all lawful means and measures." The members were elected and had to be in the business. They were also required to be "respectable citizens of the United States [who] may become an honorary member of the Union by election…and paying into the Treasurer of the Union the sum of Two Dollars…and execute a bond in the penal sum of One Hundred Dollars, with good security, to be approved by the *Union*, conditioned that he will faithfully support and maintain the Constitution." Best had made his way in America, and the Gross Park Hotel now made possible a comfortable lifestyle for Katherine and the children.

German Saloon in Chicago in 1911. One group's neighborhood hangout was another group's den of iniquity.

Best operated the Hotel above the saloon, and he had two permanent boarders at the Gross Park Hotel. The business was little more than an extension of his household. The Bests were like other saloon-owning families of late nineteenth century Chicago. Noted historian, Perry Duis describes their life, "The children performed clean-up chores and worked the basement beer pump, while the wife carried out her own important tasks…The wife of a saloon-keeper, however, often led a strenuous life.

Jacob, brother George and sister Maryanne Best, undated.

Keeping up the noon lunch was no small task, and she had to do all her family work in addition, and was often called on to help with handling drunken men. The effect on the family depended on whether the family life could be maintained independent of the saloons. In one case that we know the children were brought up with high ideals and the son never touched a drop of liquor. In other cases demoralization was inevitable, as saloons were often hangouts of disorderly characters in the neighborhood." The saloonkeeper was at peril operating a public institution morning and night. He was a businessman unpopular with reformers and all-too-popular with crooks. Not all problems were the result of rowdy customers. Fights were staged just outside the saloon door to lure the owner out and make possible a robbery inside. Or a boy might run in saying his father-waiting outside-wanted to buy beer in bottles. As soon as the owner stepped out, the boy cleared out the register. As one saloonkeeper advised, "If your regular customers depart early, better to close up and lose a few straggling nickels than have your friends say next day, 'How natural he looks, wonder who did it?'...Many good fellows have been shot down in cold blood for keeping open too long chasing the nimble nickel." Jacob Best had only his wits along with the gas lamps to protect him from those criminals who thrived in the dark.

Best managed to keep a respectable establishment. A steady stream of customers came for the German stout or, if they were new arrivals to the city, a decent bed upstairs. The saloon was a warm and affable place for people trying to make their way in what to them constituted the foreign land of America. Newcomers could belly up to the bar for a taste of the old country, and perhaps some conversation to go with it. In the words of a contemporary, "The story goes that when a German comes to America, he looks for just three things, a saloon, a church, and a singing society. When Germans drink they do not get mad, but just happy and want to sing and be friendly." Jacob Best acted as a friend.

CHICAGO AND THE CATHOLIC CHURCH

For many ethnic Chicagoans, the saloon was the most important social institution after the family and the local church. Jacob Best attended St. Boniface, and the community he discovered there became central to his family. In 1864 a group of community leaders led by Father Ferdinand Kalvelage chose a site at Cornell and Noble Streets on the Near West Side. The new church opened in 1868, a year before Jacob and Katherine's arrival.

Although St. Boniface was located several miles away, Jacob readily joined the congregation. He tackled the unpaved, often muddy, streets of Chicago to be among fellow German Catholics; churches far closer did not qualify because they were "foreign" to him, with Irish or Polish congregations. Despite the fact they were thousands of miles away from Prussia, the church reminded Jacob and Katherine of life in Hettenleidelheim, and they felt at home. They also felt safe.

Jacob had settled in a city that was dominated culturally, economically and politically by Protestant denominations. There were occasions when it seemed that the Reformation would be played out a second time along State Street, and the first Catholic was not elected mayor until the 1890s. It was no surprise then that Germans passing one another on the street identified themselves by naming their home parish.

The church history of St. Boniface notes that Father Phillip Albrecht understood that "men [and women in Chicago were] ravenously hungry for a place of worship. These men wanted a church!

His decision was quick and pronounced: 'You shall have it.'" In 1868 at a cost of $7,500 the first St. Boniface Church was erected to serve a parish population of twenty families. The bell given to St. Boniface was reportedly the one "which rang from the tower of old Fort Dearborn," the birthplace of modern Chicago.

Father Albrecht was the first clergyman appointed to serve the new church. He encouraged parish picnics and founded the St. Boniface's Unterstuetzungs-Verein, a support society to maintain and protect the widows and orphans of the parish. The German religious community of Chicago celebrated life through their picnics and their worship. The picnics at the beer gardens were a celebration of community modeled after traditions from the "old country." According to historian Perry Duis, the beer gardens "had tables and chairs rather than a bar, and the food they served was often as important as the beer in attracting trade. Music was essential, but what started as small brass bands eventually evolved into orchestras, famous soloists, and vaudeville-type performances…Fischers, [an establishment] on the shores of Lake Michigan on the northern edge of Chicago, advertised its cool breezes and bathing beach."

The beer gardens were a constant source of conflict between reformers and the German community of Chicago. To temperance advocates, they were another example of the excesses brought on by liquor. An incident at one beer garden proved the danger in all of them. But to immigrants, the gardens represented a cherished link to the old country and its traditions. This was where families went to celebrate, and didn't the reformers believe in families? Without ever intending to, Jacob Best found that his choice of business and faith in Chicago made for controversy.

On October 8, 1871, the Chicago Fire destroyed much of Chicago. St. Boniface escaped, and the congregation did everything in its power to relieve the situation. It was in such a time of crisis that Jacob and Katherine may have realized the importance of a faith community. The parish history recalls, "St. Boniface Church and School were thrown open for sufferers. Father [Clement] Venn brought clothing, bedding and vitals for their countless guests. He came with hams, bacon and sausages strung together, and he had six hams hanging over his shoulders, three in front and three behind; from his elbows hung slabs of bacon, and in his hands in front of him he held strings of sausages. His face was black with cinders and smoke, and he walked very slowly least he might drop some of his valuable cargo." Jacob did not need to share in this sooty feast. It was enough to know that a clergyman struggled to make it possible.

Best wanted a church that offered the Holy Mass in German as it provided a German cultural and religious education for his children. Not all German Chicagoans were interested in a life so firmly rooted in church. There were those who found comfort in political protest. In fact, some German-speaking members of the city's working class were among the harshest critics of American capitalism. Their outspoken views led to a backlash following the Haymarket Square Riot outside the McCormick Reaper Works in May 1886. Eight anarchists were tried and found guilty for instigating the incident that left seven policemen dead. Five of the convicted men were of German descent. To any Chicagoan quick to judge, not all political problems were rooted in the German immigrant community, but too many German immigrants seemed to be part of their city's political problems.

For immigrants such as Jacob and Katherine, St. Boniface offered the perfect place of refuge. Yet social tensions found their way into church as well. With a growing congregation, the collection of pew rents became difficult; without a steady flow of income, the church could not maintain its many social organizations. Another headache for Father Venn was the Unterstuetzungs Verein. This group seemed more interested in drunken revelry than raising funds for the poor of the parish.

Father Venn faced a two-fold threat here, both to his authority and the image of the German community. As pastor, Father Venn expected a certain level of obedience from his congregation which he found lacking. Continued unrest could lead to schism, as it did for Polish Catholics in Chicago. Just as bad were the drunken stereotypes the Verein played into. Push literally came to shove one Easter Sunday morning, and the Verein was disbanded after a tempered brawl unfolded in the aisles of the sanctuary.

Where was Jacob in all this? Like countless other immigrants, he was discovering that life in America could be difficult in ways no one could imagine. His solace, such as it was, lay in the parish-based confraternities and sodalities meant to foster the devotional life of ordinary Catholics. There were five such groups at St. Boniface for parishioners. Between Mass and club, the Best family found the will to carry on.

AN EARLY DEATH

On September 30, 1881, Best paid $27.50 for two burial plots at St. Boniface Catholic Cemetery on north Clark Street. The cemetery had been founded in 1863 by four German Catholic churches in Chicago as a place to bury the faithful. On June 10, 1886, Best's wife, Katherine died of peritonitis. Nine months later, Jacob's life was cut short at the age of 52 by cirrhosis of the liver. Best had worked as a bartender and then a tavern owner. Alcohol may have become a workplace hazard impossible for him to avoid.

Jacob and Mathilda, his second wife, in 1886.
Only one of them appears to have married out of love.

In 1887 Best had an estate valued at $5045.44. According to his will, the debts and estate settlements amounted to $4855.85. The will stated that his estate could barely afford the ice for the preservation of his body prior to burial. On October 15, 1888, his Gross Park boardinghouse/saloon, a parcel of land across the street and a 120- acre farm in Washington County Illinois were sold on the Clark Street courthouse steps. On September 7, 1886, three months after Katherine's death, Jacob married Mathilda Lange, age thirty, who received $1,590.00 of his estate. The Michael Brand Brewing Company received $2,033.31. The remainder of the estate totaled all of $189.59. It was divided equally among the four surviving children.

The Cook County Recorder of Deeds lists the Probate record proceedings of the land ownership of Jacob Best. On "December 28, 1888, Mathilda Busse (who married Herman Busse by 1888) was the Administratrix of his estate," and the land was sold for $1,860 to August, Henry and Alfred Uihlein. The record indicates that Mathilda was principally interested in getting as much of his estate, as his legal wife, as she possibly could acquire. The 1886 Lake View City Directory lists the Gross Park Hotel, Jacob Best Prop., Lincoln Av, nw corner Roscoe, Gross Park. The following year shows the Gross Park Hotel, Mrs. Jacob Best, Prop. Mathilda ultimately sold the properties to the highest bidders, remarried, and left Jacob's children to fend for themselves.

The untimely deaths of their parents left young Jacob and Maryanne orphaned at the tender ages of nine and seven, respectively. They were placed in the Catholic Angel Guardian Orphanage (now Misericordia) on Ridge Avenue. With proceeds from the operation of St. Boniface Cemetery helping support the orphanage, it seemed the logical place for young Jacob and Maryanne. The other

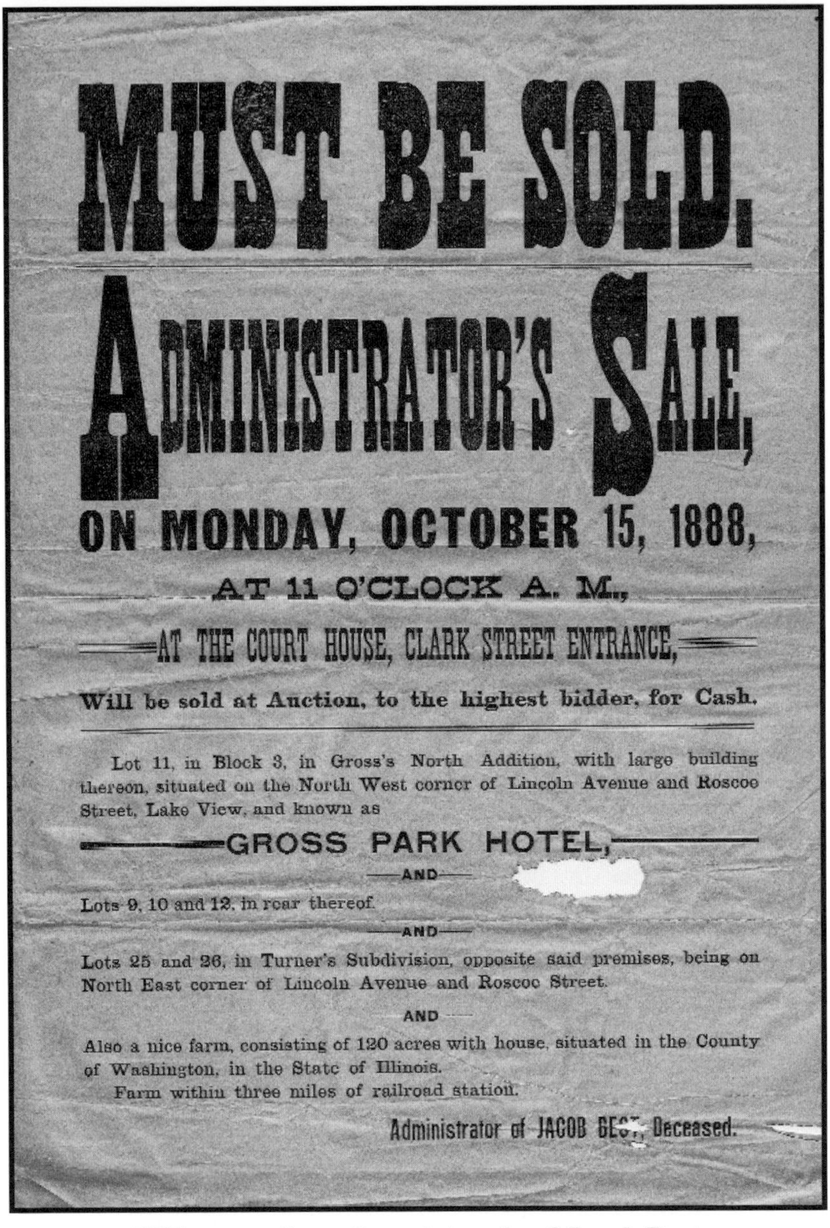

MUST BE SOLD.

ADMINISTRATOR'S SALE,

ON MONDAY, OCTOBER 15, 1888,

AT 11 O'CLOCK A. M.,

AT THE COURT HOUSE, CLARK STREET ENTRANCE,

Will be sold at Auction, to the highest bidder, for Cash.

Lot 11, in Block 3, in Gross's North Addition, with large building thereon, situated on the North West corner of Lincoln Avenue and Roscoe Street, Lake View, and known as

GROSS PARK HOTEL,

AND

Lots 9, 10 and 12, in rear thereof.

AND

Lots 25 and 26, in Turner's Subdivision, opposite said premises, being on North East corner of Lincoln Avenue and Roscoe Street.

AND

Also a nice farm, consisting of 120 acres with house, situated in the County of Washington, in the State of Illinois.

Farm within three miles of railroad station.

Administrator of JACOB BEST, Deceased.

1888 poster from the estate sale of Jacob Best.

two surviving children, Tillie and George, were older and considered ready to fend for themselves in the city. Jacob's half brother Michael Best had also settled in Chicago and he too helped to comfort and settle Jacob and Katherine's family after their parents early deaths.

An estate of $5045.44 would seem rather paltry in relation to modern times. Taking into consideration an inflationary rate of 3 percent over one hundred and fifteen years, Best's assets were worth $155,558.79 in the year 2002. In 1881 skilled craftsmen like carpenters and bakers saved perhaps $27 from their yearly earnings. The $1779 passed from Jacob Best's estate to his second wife and four children thus was the equivalent of sixty-six years of savings by such a skilled laborer. If only for a moment in death, Jacob Best was reasonably well to do compared to other German immigrants in Chicago. But that was cold comfort to his children.

JACOB BEST-FROM ELDER TO YOUNGER

Jacob Best: Here was a working-class, first-generation, German-Catholic, boardinghouse/saloon owner and member of St. Boniface Church. Few people would ever know the trials and tribulations. Such could be the outlines of his epitaph on his gravestone in the St. Boniface Cemetery on Clark Street.

Young Jacob, at age nine, was placed in Angel Guardian Orphanage in Chicago. He had to learn how to fend for himself in the orphanage. He was there for a short time that marked him for life. He was determined to become a man of influence and means.

Aunt Tillie, as Jacob referred to his older sister, helped to make his dreams a reality. She especially nurtured his desire to rise above his tavern heritage. Young Jacob Best had been raised in a difficult environment. The home on Cottage Grove and then the Gross Park Hotel were both constant reminders of his outsider status; the rough and tumble nature of his father's saloon also bore a stigma the boy felt dearly. Young Best would struggle from age thirteen to age thirty to make a name for himself in a city forever on the make and on the move. Like his father, he would become an entrepreneur. But young Jacob would not end up being a businessman by selling his customers anything to drink.

CHAPTER 2

CHICAGO, 1878-1908:
The Development of a Second Jacob

THE NEXT (JACOB) BEST

Between 1871 and 1900 Chicago grew almost exponentially. "People began calling it America's city," writes urban historian Donald Miller, "the concentrated essence of Americanism." From its beginnings as a fur-trading outpost at the southern end of Lake Michigan, Chicago has always been a place where the locals have wanted to make money, and lots of it. The atmosphere was high capitalist, even crassly so, and not everyone approved. In 1889 while on a tour of Chicago, Rudyard Kipling remarked of money-mad Chicago that "while people hurried along, they talked money through their noses." In the midst of such a bustling city, immigrants may have wondered what the future would bring.

In 1878 the third Best child, a boy, was born in the home on Cottage Grove. His father died when he was nine, in the same year that Marshall Field retail store was completed, a masterpiece in commercial Romanesque. The timing may have been a coincidence, or an omen. This boy would have something to do with the local business world.

Commerce made Chicago the architectural capital of the world, with a healthy assist from the likes of Louis Sullivan, William Le Baron Jenney, John Wellborn Root, Daniel Hudson Burnham and others. In the last decades of the nineteenth century, architects and engineers hammered out the design for that most American of buildings, the skyscraper. The roots of the Chrysler Building extended further west than any New Yorker cared to imagine.

Steel-beam construction may have been discovered purely by accident, historian Donald Miller suggests, "Leaving work on the Home Insurance Building early one afternoon, Jenney surprised his wife, who, rising to greet him, put a heavy book she was reading on top of a bird cage that was sitting on a table by her chair. At that an inspiration struck Jenney…if so frail a frame of wire would sustain so great a weight without yielding, would not a cage of iron or steel serve as a frame for a building?" This simple principle soon revolutionized urban life.

Chicago became a city of steel, and steel was the supreme structural element of the industrial age. However, steel did not create itself but was the product of a process dependent on the energy produced from coal. In that year 1886, Jenney also designed the Home Insurance Building, the world's first skyscraper. Young Jacob did not know it at the time, but he would grow prosperous in a business that helped fuel the creation of such a marvel of technology.

If Chicago was known for architecture, it also became a model urban transportation center. Mass transit arrived in 1859 with the introduction of horse-drawn streetcars on city streets. But horsepower was both unpredictable and prone to "apples" that littered the pavement. Horses gave way to cable cars which in turn yielded to the electric streetcar by the turn of the century. In addition, the Columbian Exposition of 1893 demanded a transportation system capable of handling the hordes of visitors who descended on the fair site daily. Thus was born the Chicago "el" or elevated line, from Twelfth Street to Jackson Park. The Father of Chicago mass transit was Charles Tyson Yerkes, who built streetcar and elevated lines wherever he smelled a profit to be made. Yerkes even "looped" downtown with "el" tracks, giving downtown Chicago a nickname (and noise) still heard today.

The trains, like the rest of Chicago's economy, ran on coal, not that young Jacob cared in the wake of his father's death. But, if the child is at all an extension of the parent, the young Best inherited something of incalculable worth from his father: As Jacob Best worked and strived to succeed, so would the next Jacob Best.

A YOUNG BOY IN THE CITY

Tillie removed her siblings Maryanne and Jacob from Angel Guardian before too long. For six years she clothed, fed and provided them shelter in her home on 265 N. Lincoln Avenue in Lake View. Young Jacob attended school through grade six. Although Best never went to college, as an adult he attended one of the first night schools in the United States. Decades later, Best said that, when he realized his education was not equal to his ambitions, he enrolled in night school at the LaSalle Street YMCA to secure "a better grasp on commercial affairs."

The YMCA was the immigrants' university. According to Perry Duis, the students "were a diverse lot that included older immi-

Young Jacob Best on the streets of Chicago.

grants struggling to master English, elderly African Americans finally getting a chance to learn how to read and write, women taking home economics courses to improve their households, and ambitious young men and women trying to absorb the 'high tech' of their era: office machines, telephone 'instruments,' automobiles, and electrical equipment. Their ages ranged from fourteen to eighty-five, and they came from forty-four nations, but they were all part of one of Chicago's greatest achievements, mass education through night schools." What the city offered, immigrant and working-class students knew to take advantage of through education.

Tillie exerted her influence on Jacob in at least one important area, his faith life. She had married Charles E. Simon in 1886, a Lutheran and with her encouragement Jacob became Lutheran. Together, brother and sister attended St. Pauls German Lutheran

Church on LaSalle Street just north of downtown. In 1896 the church relocated further north, to Fullerton Avenue in Lake View.

Faith is that most personal of human beliefs. It allows people to approach God in a manner that best suits them. At the same time, it makes the individual part of something greater. In becoming Lutheran, Jacob placed himself in two communities, the one ethnic and the other metropolitan. As a Protestant, he lost no standing among Germans, who were perhaps evenly divided between the Lutheran and Catholic faiths. And, as a Protestant, Jacob looked to be more mainstream American, at least to WASP Chicago.

Three great Chicago events-a conflagration, a riot and a celebration-likely shaped some aspects of Jacob's personality and outlook on life. Although he would not be born until seven years after the tragedy, Jacob could not have avoided the public memory associated with the Chicago Fire. It was a catastrophe of incredible dimension, totaling some $190 million in losses. Newspapers reported that 17,450 buildings burned to the ground and almost 100,000 persons were left homeless. The death count of 250 victims is more guess than estimate. A swath of destruction traveled north and west from Mrs. O'Leary's through downtown up to Fullerton Avenue. Photographs show a landscape not unlike that of Hiroshima following the explosion of the atomic bomb.

Chicago did not lack for critics on the eve of the Fire. Some observers saw it as divine retribution, either for the city's vice district or its politics. One Indiana newspaper with Southern sympathies during the Civil War noted how the Union Army had caused greater destruction. Troops hailing from Chicago did their "full share in the destruction of the South. God adjusts balances. Maybe with Chicago the books are now squared." If anything, such attitudes only fueled the desire to rebuild.

The city overcame disaster through a commitment to technology and talent. The skyscraper and the catalogue were among the post- Fire innovations that returned Chicago to prosperity. It did not matter what William Le Baron Jenney or Montgomery Ward did before they came to Chicago. What mattered was the talent they brought with them. The city had an insatiable appetite for that commodity. It was something that would work to Jacob Best's advantage in business.

As the child of German immigrants, Jacob could not have escaped the touch of Haymarket. The riot fed a longstanding fear of newcomers, and it highlighted the danger that came with the decision to ignore or reject the core beliefs of American citizenship. If in the years after 1886 young Jacob Best heard a German joke, he likely was experiencing some not-too-funny fallout from Haymarket.

The Chicago press attributed the growth of German working-class militancy in the 1880s to the arrival of revolutionaries driven out of the fatherland by Bismarck's sweeping anti-socialist laws. But conditions in Chicago were most responsible for the rise of German-American radicalism. Germans were nineteenth-century Chicago's, and nineteenth century America's, largest ethnic group. Their proportion of the city's population never dropped below twenty-five percent between 1840 and World War I. They settled in tightly concentrated enclaves close to industrial districts. Rich and poor, Catholic and Lutheran, Bavarian and Prussian, they lived together in virtually self-sufficient urban villages. For anyone who equated America with notions of a New England town, Lake View was a frightening place-and labor protest was a most frightening act.

It was never proven who threw the bomb that May 4th resulting in the deaths of seven police officers. The eight men found guilty of murder had distinctly German last names as Engel, Fischer and Schwab. An aura of suspicion settled upon German Chicago. It lessened over decades only to return in 1914. Through no fault of his own, Jacob Best found himself under suspicion. His response would be to demonstrate an Americanism no one ever called into question.

If Chicago bestowed its favor on those who demonstrated talent and patriotism, the city also put on a good show for everyone from time to time. There may have been no better show during Jacob Best's life than the Columbian Exposition of 1893, when the world came to Chicago for its fair. The fair drew an estimated 27 million people, 14 million from outside the United States. Donald Miller reports that, on opening day, "President Grover Cleveland, Mayor Harrison, Daniel Burnham, and Bertha Palmer, the queen of the occasion, arrived in open carriages just before noon. They were greeted by the cheers of a crowd of 200,000 persons packed together in front of the speaker's stand at the Administration Building. 'Buffalo Bill' Cody, in town for the summer with his Wild West Show, waved his immense white hat from his prominent position in front of the crowd, while Jane Addams, caught somewhere in the crush, felt someone snatch her purse."

Aunt Tillie may have brought Jacob to the fair, or he may have heard about the "White City" designed by Daniel Burnham from a friend or neighbor. Where to start? There was the Children's Building located between The Woman's and The Horticultural Building. The Official Guidebook describes it as "Generally plain, beautiful though simple; two stories high, with a roof as a playground. Everything to instruct or amuse children is found in this building. Children may be left here by the hour or by the day in charge of careful nurses." Or brother and sister may have spent a few minutes in the White Star Building, erected by the White Star Steamship Company, later the proud owner of the Titanic.

By all means, they would have wanted to visit the German Pavilion with its 5,000 exhibitors coming from 230 cities and towns. The Official Guidebook says, "The Krupp gun exhibit in itself is one of the marvels of the Exposition. On the Plaisance is a German village and Hagenbeck's zoological arena from Berlin. The German building is one of the most beautiful of the foreign structures, and her pavilion in the Manufactures building is unsurpassed." Visitors from Lake View were tempted by a tug of pride in the old country. If Tillie took Jacob on the 250-foot high Ferris Wheel, the first of its kind in the world, it would have been thrill enough.

Young Jacob Best was but the orphaned son of immigrants, and there were so many others like him. But for an ambitious child, turn-of-the-century Chicago may have been the best place in all the world to live in.

FIRST JOBS, BEST'S JOBS, CHICAGO JOBS

Jacob began working at age thirteen, though he did not discover his entrepreneurial talent until he was thirty years old. Luckily, Best did as meatpacker Philip Armour preached, "Always keep at it. Don't let up. Let liquor alone, pay your bills, marry a good wife and pound away at whatever you want and sooner or later you'll make good." What worked in the Stock Yards worked in the coal business, too.

Best worked hard and held several positions from 1891 to 1909: clerk in a law office, lithographer's apprentice on South Water Street, tea-and-coffee store clerk, laundry wagon driver, Marshall Field & Company retail fur clerk, grocery peddler and store owner.

On November 1, 1897, the owner of Charles Wolff, Wholesale And Retail, Coffee, Tea, Spices And Fine Specialties, wrote a general letter of recommendation for his nineteen-year old employee:

> *To Whom It May Concern:*
>
> *The bearer Mr. Jack Best has been employed by me as a clerk and collector for the past two years and has faithfully discharged all the duties developing upon him. He was very functional at his work and courteous to customers and reliable in his accounts. He is not addicted to [hard] liquors and bears a good moral character.*
>
> *I wish him every success*
> *Very respectfully*
> *Charles Wolff*

There was something about this young man that bore watching.

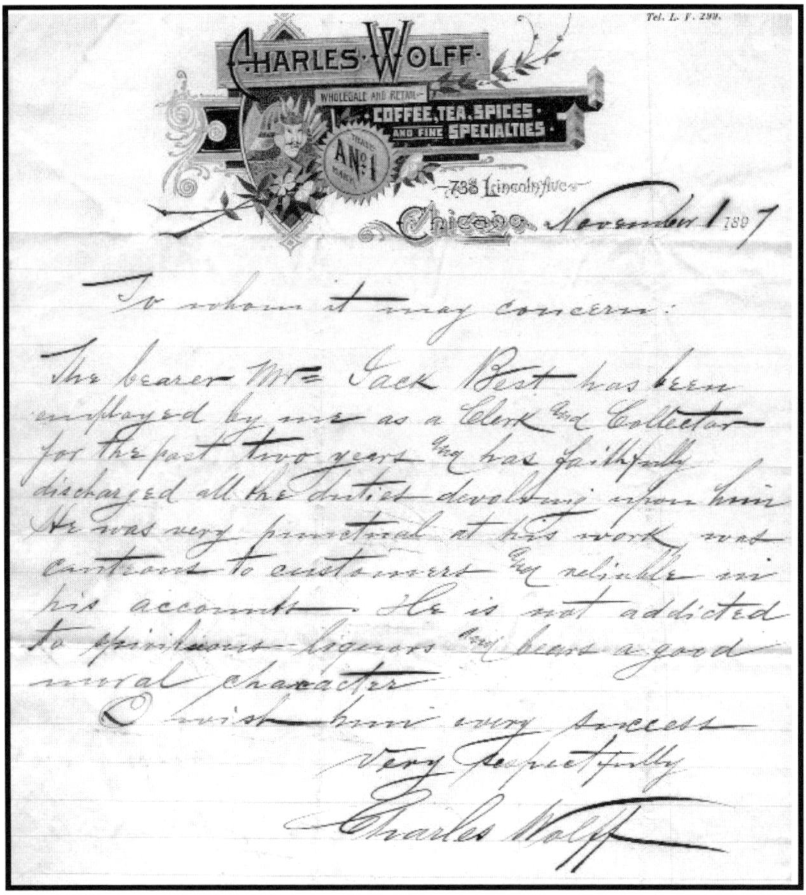

CHARLES WOLFF
WHOLESALE AND RETAIL
COFFEE, TEA, SPICES
AND FINE SPECIALTIES
TRADE AN°1 MARK
738 Lincoln Ave
Chicago, November 1 1897

To whom it may concern:

The bearer Mr. Jack Best has been employed by me as a Clerk and Collector for the past two years and has faithfully discharged all the duties devolving upon him. He was very punctual at his work, was courteous to customers and reliable in his accounts. He is not addicted to spirituous liquors and bears a good moral character.

I wish him every success.

Very Respectfully

Charles Wolff

Charles Wolff employment letter, 1897. Most letters of introduction in Chicago tended to be more political.

MARSHALL FIELD & COMPANY AND FIRST LOVE

On April 3, 1899, Jacob left his laundry wagon for the promise of a steady salary. He entered the prestigious Marshall Field & Company store in the Loop to apply for a sales job. The city's premier retailer hired him for a clerk's position in the fur department. It was a position that brought the twenty-one year old in touch with the smart set of a growing city.

At Marshall Field's a customer could buy almost anything, and Field's was a business that would go almost anywhere for a customer, even to New York. Jacob saved an article from *Cloaks and Furs*, a leading New York trade-paper, showing how aggressive his employer could be:

> *To the average Easterner the firm of Marshall Field & Co. is known simply as a large buyer of all sorts of merchandise. Now that this typical American firm has found a foothold in the New York market, it will rest with the vast number of buyers to decide whether the firm's venture as manufacturers of two of the most difficult articles that can be made, is well conceived. Whether its execution is to be successful or not, Marshall Field & Co., having acquired the best talent for the purpose, has certainly succeeded in turning out as perfect a line of fine and medium furs as can be produced on this globe.*

Field's Furs, 1900.
(Courtesy Marshall Field's)

No one knows for sure how Jacob met Hulda Hammerstroem. But enough evidence survives to suggest the relationship grew with bits of self-promotion, the formality of courtship, romance, and persistence. Jacob wrote Hulda relentlessly just as he took her out for dinners, boat rides, dances and plays. The courtship lasted from 1899 through 1904, while Best worked at Field's. Hulda was from a German family whose parents had met in Chicago.

William and Minnie Hammerstroem married on August 5, 1877, at St. Pauls German Lutheran Church in Chicago. They lived at 333 N. Burling Street and had emigrated from Germany to Chicago in the mid-to-late nineteenth century. Hulda's father operated William Hammerstroem Coal Company at two locations, 632

Clybourn Avenue and 98 Willow Street in Chicago. The company was a wholesale/retail supplier of coal and wood.

On September 3, 1900, Jacob wrote an introductory letter to Hulda. There was barest hint of poetry and romance in it. Instead, the clerk informed the customer he met at the counter about the fur business at Marshall Field & Company:

> *I here send you a souvenir pamphlet of the store for which you have expressed a desire, the circular therein gives different views of our fur factory which might be of interest as it gives the principal sections of the factory and shows the process each garment goes through from skin to garment. It may not be plain to you as the cuts are rather small, but should I ever have the opportunity I would be pleased to explain it to you, and I am sure it would be very interesting. I am quite interested in this circular as the views were taken while I was employed in that department and as I am quite prominent in one of the cuts. I wonder if you could find me.*
>
> *The ladies at the store were angry with me this morning for not giving them an introduction to you, and said they did not think I was so selfish, but to be candid about it, I didn't think you would like to be bothered by being introduced to everyone.*

Your sincere friend, Jacob Best

And so began courtship. A short two days later, Hulda responded on letterhead from her father's coal company:

> *J. Best*
>
> *I am very sorry to hear that all the young ladies at the store were angry with you on my account, that is, I didn't think that they were so much interested in me as to desire an introduction, properly speaking you were not to blame, why didn't they ask you in the first place and then to tell you when 'it is all over now.' I hope they have all made up with you again.*
>
> *Your sincere friend, Hulda*

This chance encounter proved to be the best thing about Jacob's job at Field's, if not the best thing in his life. Young Best was a hardworking man frustrated over his position as a retail clerk. He thought he had improved his station in life by leaving street peddling behind for a position with a steady salary. However, a regular income encouraged a complacency in others that Jacob wanted to avoid. In his letter of Oct. 30, 1901, he wrote:

> *Dear Friend Hulda:*
>
> *Now about my resignation, I will write and send it tonight but as for writing Miss Boddiker, I don't think I will...Had splendid success catching the train Tuesday night, and had lots of company too-that is, there were about thirty Negro's in the car, talk about rag time, well I guess yes.*

While Jacob did not resign as promised, his aspirations were set on becoming something more than a retail clerk. He was frustrated despite the fact that he was earning a respectable wage. By contrast, a 1911 study of Packingtown (the community around Chicago's Union Stock Yards) workers showed that male heads of households brought home on average just over $9.50; this was almost six dollars short of what the study calculated was necessary to maintain a family of five above the poverty level. Jacob sensed that remaining single kept him from being poor while marriage could make him that way.

Jacob's precarious finances soon spilled over into his personal life and his relationship with Hulda. On January 12, 1902, he let slip his anxieties over their difference in status:

> *No you did not offend me, and I probably owe you an apology for answering you so abruptly and being (as I fancy I was) disagreeable all evening, but my position has annoyed me so lately that I have been exceedingly cross, so I hope you will kindly excuse me. In reference to attending theatre with you, I am very sorry to say that financially I am not in a position to make such arrangements for the present. Have shamefully neglected my savings account lately, and am very anxious to make up the deficiency. My duties at the store do not make it necessary for me to stay evenings now, as business is very quiet.*

Miss Hulda Hammerstroem.

Jacob was now bored with his work at Marshall Field's and the limited opportunities it promised him. He had found his way into the good graces of one of Chicago's better German families, yet he did not believe that he could live up to their standards. How could a simple retail clerk aspire to see a lady of such obvious refinement? And so Jacob worried. Ultimately, the young man saw greatness in his future. The challenge would be convincing Hulda of it.

Unfortunately for Jacob, 1902 was not the easiest year to be a second-generation German version of Horatio Alger. On July 15, Hulda received a letter from a friend commenting on labor unrest in Chicago, “The latest talk of the City is ‘Strike.’ Everybody is either striking or talking about it. The Commission Business has been affected very badly by it, because the teamsters can not haul the goods from the docks or freight houses, their lives are in danger. Everything which comes into Chicago is left at the freight houses, and is going to ruin. No doubt your father also notices the affect the strike is taking to his business.”

He did. Hulda’s brother George was working in the coal business when he wrote her three days later, “We have had a big strike here again which was settled yesterday, and which was a loss to R.R. and wages to men and loss to merchants in the sum of ten million dollars.” If Jacob were to make a name for himself to impress Hulda, it would have to be as something other than a union local president.

Whether or not he knew it, Jacob was witnessing the growth of modern industrial America simply by watching Hulda and her family. That abstraction known as the “trust” was not something John D. Rockefeller played with in Cleveland. Hulda’s first cousin Minnie Thilo wrote all about it on February 12, 1903. The letter described the faith that her family had placed in the trust, a controversial forerunner of today’s conglomerate, “Papa formed a trust or rather a combine taking in these two plants and has control over three now, this is a great deal for him…These two plants are going to be a very great help to us, as we can get our products in from the East as far as K.C. for a much less freight rate, than to Topeka, and we can grow pickles in Colorado and have them put down with less current expense than to buy from Wisconsin and Chicago, so you see how this will help us in the future."

Transportation was essential to the growth of Chicago. The railroad industry made it the gateway to the West, and commerce flourished. The railroad conquered both time and distance, to the point that "the lobster died in Cleveland," as Perry Duis describes the fate of one selection on Chicago menus in the 1840s, no longer held true. The lobster lived, if only to meet its demise in a restaurant somewhere beyond the Ohio Valley. Or, in the case of Mr. Thilo, the railroad drove down the cost of pickles by creating a truly national market, and what held for pickles did for just about every item imaginable. Somewhere in all this lay Jacob Best's future.

And so did love. By the late summer of 1900, Jacob felt confident enough to invite Hulda to a picnic. He wrote her, "Enclosed you will please find an invitation for a Basket Picnic. You have undoubtedly heard it before, as your aunt and uncle are quite interested as to its success. There will be quite a large group and most of all very nice people, and I feel quite sure that if you will go you will have a pleasant time." The invitation was done in a somewhat awkward mix of German and English:

Vater schaft nicht on Lontag
Yourself and family are cordially invited
to attend the Second Annual Private Basket Picnic
given by "The Owls"
Sunday, August the nineteenth, nineteen hundred
at Anton Kill's Grove, 2544 Lincoln Avenue
Rain or Shine
Ceremonies open at ten a.m. sharp
Take Bowmanville cars to the end

The courtship of Jacob and Hulda was a very formal affair, conducted mostly through correspondence (delivered, no doubt, too slow to keep young lovers happy) or in the presence of chaperones. Usually accompanying Jacob on dates was his younger sister. “Maryanne begs to be excused but asks me to say that she would be glad to have you call,” wrote Jacob in typical fashion. “However, if it will be agreeable to you to go alone with me, I will meet you as suggested at 4 PM at Diversey Street.” It was indeed a different time, if not world.

Passion existed between the lines, as when Hulda wrote Jacob on September 3, 1900:

> *“Words could not express my feelings when I read those few lines of your note, which meant so much to me and perhaps you would not have sent them had you reconsidered and not been so hasty…Indeed, I regret to say, had I known that my going away would be the cause of all this unfortunate happening. I would have gladly stayed at home…so I will ask you to call and spend Wednesday evening with me, for I know I will feel greatly relieved if I can speak to you in person and perhaps by then you will feel more agreeable towards me.”*

One of Jacob and Hulda’s favorite places was the famous Philip Henrici Restaurant at 67 West Randolph Street. This fine German establishment opened in 1868 and rebuilt after the Chicago Fire. Jacob kept a souvenir from 1894 commemorating the restaurant’s twenty-fifth anniversary. Mr. Henrici wanted his customers to know, "It has been my intention to make this establishment, where I have been located since July 1894 one of the most artistic and cozy of places, in keeping with all the progress of modern times. No expenses of effort has been spared to place the same before the public as one of the most popular Restaurants and Café's in the country. Being close to all lines of cable cars, our location is accessible to any for all parts of the city, and is within one square of the City Hall and County Court Houses, and is directly opposite the Schiller Theater."

The lavish interior, with a seating capacity for five-hundred diners, was decorated by architect August Fiedler so as to be "Fit For The Fittest." Jacob and Hulda frequented Henrici's over the course of their four-year courtship. It was both uniquely German and, as the proprietor promised, "Cozy," a feature appealing to any young couple.

On July 28, 1901, Jacob invited Hulda to take a Sunday cruise across Lake Michigan. "It will be 'Lovers Lane' tomorrow evening," he promised. Of special interest was the fact that the chaperone had an upcoming engagement "so you will please excuse her. As for photos, she has disposed of them some time ago, and I haven't had any taken. But how about yours, aren't you going to give me one?" The two lovers took boat rides, went to picnics together and corresponded back and forth. They even attended formal dancing academies in the city. Jacob wrote Hulda on November 3, 1901, "Indeed, I did think about the dance, not so much the dance, but …"

Advertisement card for William Hammerstroem & Co., coke and wood. It made smart advertising for an oftentimes dirty business to appropriate images of nature.

Like many other young couples in love, Jacob and Hulda did have their moments of misunderstanding. Jacob wrote on March 12, 1902, about one of them, "Answering yours of the 2nd, will say that I am pleased to note that you have favorably considered that little dispute of ours. I thank you for your kind consideration and assure you that neither subject will ever again be mentioned by me." But the subject did come up again one month later. Jacob was a proud German who bristled when Hulda referred to his interest in her as "begging."

On April 9, he wrote her again, "But Hulda, I do want to say that I hope you will never again use that word 'beg,' for I detest it. You know that I would do most anything to please you, but I must ask that you try to be reasonable. You know my reason for not going, you said you had an escort. I am sure he is agreeable or you would not have accepted him. I am willing to sacrifice the good time I know I would have had because I think my cause is a just one. Now if you say why I should go, I will cheerfully reconsider it, that is, if it is reasonable."

The misunderstanding continued to the point that, by May of 1902, the relationship seemed doomed. Jacob wrote:

> *I have so frequently and in so many different ways thought of our conversation of Wednesday evening that I have become quite perplexed. I can't remember whether I promised to write, if I would call, or, if I decided not to. Although it is extremely unpleasant for me to write it, I must say that I have not changed my mind. I have wondered, Hulda, whether you have thought about it at all, and if you have, whether you do not think just as I do. However, may I ask the pleasure of receiving an occasional letter, it will be a delight for me, and I hope not an awful task for you. Hoping that I have not annoyed you, and that I may always have the honor of your esteemed friendship, I remain with kindest regards.*

Seven days after sending Hulda a letter that suggested they were no longer a couple, Jacob wrote on May 30, 1902, "Was sincerely pleased, and I might add surprised to receive your ever welcome letter last evening, for I was just beginning to think that you would not write. Please do not mention that conversation again,

Hulda. I know it doesn't amount to more than a jest to you, but to me it was a painful obligation, and now it is an unpleasant recollection. If you think different from what I do, I am sorry but hope that whatever your thoughts may be, they are not disrespectful."

By July the relationship seemed to be over. Jacob wrote that:

> *I was very sorry to observe the sarcasm and formality your last epistle suggested. Why did you write so?*
>
> *I have always enjoyed your letters so much Hulda, and you surely know my feelings toward you. Can't we be friendly, without being intimate, knowing the latter is out of the question? You mention "that the change will be about the best cure," I did not know you were ill, what has been the trouble? Had a very quiet 4th, but had a pleasant time Saturday at Lake Geneva with Maryanne…Was at the concert in the park yesterday and met Mr. Geo. Ahrens and also that young lady friend of yours who lives near you. I have forgotten her name.*

Over the summer Hulda traveled to Topeka, Kansas to see her cousin Minnie who was studying to become an osteopathic nurse. Jacob still wrote, but the relationship appeared strained. Then, in a leap of romantic faith, Jacob sent Hulda a gift complete with bearer. He wrote on July 25, 1902, "Have referred Miss Dorothy Vernon with a telescope to you today. I am sure you will enjoy her company on your return trip and the telescope I trust will not be a burden either. I expected to hear from you but suppose you are having such a fine time that you do not get time even to think of Chicago."

By December of 1902, the relationship was still strained, and Jacob wrote, "However, you know how I feel about accepting an invitation from her and as I feel quite sure you would rather go with other company, I beg to be excused. Kindly thank Miss Nopesise for me for the invitation, when you see her." Jacob wrote again on February 25th, 1903. It was a rather somber letter, as if a final one:

> *I am pleased to learn that your father is well again, and that you have time to write…Let me say Hulda, that I love you with my heart and soul, and that your company is more pleasant and more agreeable to me than anything on earth. You have told me that you could never think as much of me, so I must try to see as little of you as possible, for the more I see of you, the more I think of you, if that is possible. I extremely regret but think it best to ask you not to write any more, but would like to ask if I might send you my compliments January 1st and Easter Sunday. If I do not receive a reply to the contrary, I will take it for granted that this is permissible. Assuring you that I will always think of you with esteem love and friendship, and hoping that you will not think unkindly of me, I beg to remain, As ever, Jack*

But the dance continued. Jacob received a promising letter from Hulda, and he immediately responded on February 28, 1903, "Oh how nice it is to say this and to know I dare. Your letter just received has made me the happiest man in the world, and I can't wait until tonight to tell you so. Don't be unhappy when I am so joyous. Your happy and devoted, Jack."

By the summer of 1903 the couple was again together. "Happy, well I dare say, yes," Jacob wrote, "Such a dear letter and the prospect of (2) two kisses from my sweetheart, the best, dearest, and prettiest girl in the world, how could I resist. I'll be there and on time too, I am in haste. Yours devotedly, Jack. Don't forget (2)"

Hulda traveled to St. Louis and the 1903 World's Fair, Jacob wrote her on August 26:

> *Do you know it seems an age since you went away, and it is only 19 days. I do hope it is a fact that within the next 10 days you will be back again. How did you get along at the Fair Grounds, is St. Louis anything like the enclosed clipping?...I do hope my little sweetheart is in as good spirits as ever, if not, why not come home now? I am sure the pleasure it will give all, will show you and make you happy, and as you mention, 'There's no place like home.'...Please write soon, be a good girl and take good care of my dear girl. Kind regard from all, and love from, Yours devotedly, Jack.*

By September Jacob was professing his love for Hulda:

> *You know I love you more than any one, or anything in the world.*
>
> *To cause you annoyance or disappointment worries me and makes me feel ungrateful. You have always, or most always, been so very considerate and kind to me that I will always be under obligations to you…You deserve only kindness from me, and I repent that I have been unreasonable, but as sure as I live, so sure I have always tried to be kind to you, and no one knows how sorry I am that I have not been successful…'With Love, your Hulda' how happy this closing made me the first time I received it and how unhappy it makes me tonight to realize that it is only written. Yours devotedly, Jack*

Hulda's cousin Minnie wrote Hulda from Topeka on January 18, 1904. It was a case of good news traveling fast:

> *My dear Hulda, you do not know how surprised we all were to hear of your being engaged!*
>
> *I was so happy to think that you had the X-mas tide of your life. Indeed - I know you are the happiest girl in the universe, and to Jack - well well, dear Hulda you have my dearest and best wishes for a prosperous, happy and joyful future. And Jack loving and worshipping you as he does, should cer-*

> *tainly be the happiest and proudest man that breathes, as his choice is certainly fine and you are worthy of a good, kind and loving husband…I hope that I will have the very great pleasure of seeing you stepping from your single blessedness into your double happiness. Hulda, I hope that the next time we have the great pleasure of meeting you that you would have the simple "Miss" changed into the dignified "Mrs." What does or did your parents think of your rash steps, honey? I don't suppose the shock was so very great to them, however, as they no doubt expected it, Hulda. Oh, Hulda, when a young girl stays home to learn cooking and "dress-making" there is something behind it, when ever I hear of anyone doing anything like that, I feel sure something is going to happen…I know I haven't seen my mate, as I am destined in my own minds eye to be a "spinster." However, (this is awful and worries me so) my "Astrological Reader" told mama about a week ago that she had one daughter and that she was going to be married in "1905." Now wouldn't this worry most anyone, when there is no one, practically no one at all, in view?*

On the 7th of June, 1904, the courtship ended at St Pauls Church as Jacob Best married Hulda Ida Hammerstroem.

Jacob and Hulda on their wedding day, 1904.

DANCING AND COURTSHIP IN CHICAGO

Dancing stood at the center of Jacob and Hulda's courtship. Historian Perry Duis has noted that the art form came with the first settlers, "As early as 1834 out-of-town dance teachers began including Chicago on regular seasonal circuits, and by the end of the decade there was healthy competition among several resident instructors. James A. Marshall, for instance, invited potential pupils to a one-dollar sample evening of dance at the New York House, hoping to sign them up for three months of weekly lessons for ten dollars. During the 1840s many of the schools added juvenile classes. In 1850 'A. Woods' boasted that he had traveled East to learn 'the latest style of dancing, waltzing, and galopading…together with the new and improved polka, Scottisch, Redowa, Chacuca, German Celarious, Two and Four Step Waltzes, also the German Cotillion.'"

Wealthy Chicagoans took pride in their dancing etiquette and style. Bornique and Martine dance instruction was all the rage in Chicago; to be enrolled in such classes was a sure sign of status and culture. Although some critics still questioned the morality of any form of dancing, good dance instructors did not need to fear going broke. While many churches had considered holding hands a sinful expression that led to inappropriate bodily contact on the dance floor, the formal dance card removed a good deal of the stigma associated with such activity.

Jacob and Hulda danced away long evenings at Prof. Kretlow's Dancing Academy at 401 Webster near Lincoln Avenue in Lake

View. In the notorious First Ward Ball, prostitutes joined pimps and voyeurs in an annual event that shocked the rest of the city. There would be none of that for Hulda and Jacob. Hulda wrote on February 2, 1900 to inform her beau, “Most assuredly, I shall attend Kretlow’s Mask Ball without fail, and I suppose you and Maryanne will mask?” That fall, she wrote again inquiring about Jacob’s possible attendance at the Thanksgiving Party at Kretlow’s, “Enclosed you will find an invitation for the Thanksgiving Party given at Kretlow’s, and I thought probably you and Maryanne should like to attend. As I have not heard from you for some time, I thought you must be angry with me, but I hope such is not the case.”

The following invitation describes the event:

Your presence, with friends is desired to
attend the Thanksgiving Party.
Thanksgiving Party of
Prof. Kretlow’s Dancing Academy
401-403 Webster near Lincoln Avenue
Thursday, November 29th, 1900.
Entrée at 9 o’clock Carriage at 2.
Admission 50 cents a person or 75 cents a couple.
This invitation is good for all Sociables on Thursday during the season.

Also included in the invitation was a list of upcoming events:

Special Entertainments.	
Thursday December 6th.	*German*
Thursday December 13th.	*Waltz Reception*
Thursday December 20th.	*Popular Programme*
Thursday December 27th.	*German*
Monday December 31st.	*Closing Party of the first term.*

Thursday January 3rd, 1901. Waltz Reception and beginning of the second term of this season.

The German at Kretlow's Dancing Academy was a favorite dance for the couple. Jacob and Hulda exchanged a series of letters about it. On December 12, 1900, Hulda wrote that, "The last letter I assure you I have answered, and I think you will probably recall it as it referred to the German." Jacob responded three days later:

> *Maryanne and I were at Kretlow's a week ago Thursday to a German, we expected to see you, but, I am sorry to say, we were disappointed. How are you, are you going to many dances? I dance so seldom that I fear I am forgetting it entirely. I am trying to be real good now for you know Christmas is coming, and I want my stocking filled. You ought to hear my little nieces talk about Christmas, it's nothing but Christmas all day long.*

Jacob Best (left) and his brother George enjoy a cigar in the company of their brothers-in law Max Goltz and Charles Simon. At the far right is William Hammerstroem, Hulda's father.

Hulda wrote again on December 20:

> *But as it was the Bazaar of the Church just happened during that week, and of course as I was in charge of one of the booths, I was obliged to be there. As there will be a German on the 27th. I shall have to go as I haven't been down for the last three weeks and, I think I shall need a little practice oth-*

erwise, I won't know how to Waltz. I suppose you are quite busy now during the Holidays and you have longer hours too. This weather is entirely too mild for our line of business, indeed I get plenty of spare time now - to read a little now and then. Wishing you a very 'Merry Christmas.'

The dance that got away seems to be the subject of a series of letters starting February 8, 1901, when Jacob asked:

How did you feel today after all that fun and excitement last evening? But what on earth possessed you to leave so early? If I dared, I would like to say that you are a naughty, naughty girl to disappoint me by not giving me one of the two dances I thought I was going to have with you...I have resolved you shall not escape me next time, that is, if ever I may have the pleasure of meeting you again where there is dancing, for I will ask you for an early Waltz, or Two Step, before you get your Program. Maryanne and I left after the eleventh dance, there was quite a crowd there then yet.

Four days later he wrote Hulda again:

Your letter of Sunday received last evening with sincere joy but was very sorry to learn that you were ill but hope that you are feeling better again.

Certainly you are excused for disappointing me under such circumstances but really, if you disappoint me again, I shall be inclined to think you don't want to dance with me. Indeed I did have a pleasant time, but I must confess that I felt somewhat vexed and less happy when I noticed that you had gone…Maryanne has a picture for you and we both thank you very kindly for the kind invitation to call Thursday evening, which we accept with pleasure…Yes, I was skating Sunday afternoon. I regret to say that I did not enjoy it very much, as it is rather tiresome when one is alone. I wonder if you will permit me to call for you sometime. Maryanne and I are going to the North End Refectory Friday evening, to a reception given by The Arcadian Literary Society, we expect to have a good time. I wish you were going too.

This invitation was sent to Hulda for a Kretlow's Dancing Academy May Party, sponsored by the Cordelias. After attending, Jacob reflected on the evening and said, "Had a good time at Kretlow's and it was almost 5 o'clock when I got home."

The Presence Of Yourself And Friends Is
Earnestly Desired At The
First Annual May Party
Given By
The Cordelias

Wednesday Evening, May 14th 1902
At Kretlow's Dancing Academy
401-403 Webster Avenue
Grand Entrée At Nine Carriage At One
Tickets 75 Cents A Couple

CLUBS AND CARS IN CHICAGO

In that time before television and computer games, young Chicagoans spent much of their free time as members of private clubs. For Jacob, it was only natural to join the Germanic Club on Clark Street. Jacob wrote Hulda on June 30, 1901, "I am pleased to know that you approve of my joining the Club, but I fear that when you find out how sober and quiet I am, you will be sorry that I was ever admitted to join." In truth, Jacob hoped that both Hulda and her family would be impressed. Depending on their membership, clubs brought together men of standing in politics and business. If nothing else, Jacob hoped to impress the Hammerstroems with some stature by association. Jacob also belonged to the Webster Club, although it failed to provide the kind of social activities for the couple that he wanted.

Jacob and Hulda dated at a time when automobiles were still a curiosity and an expensive one at that. In 1903 there were only 1,461 drivers registered in the city. When Jacob wrote Hulda that August, he used his brother George in yet another effort to appear the prosperous suitor:

> *Do you realize that this is the third consecutive Sunday that I have not seen you, and dreary Sundays they do seem. Even last Sunday, although miles and miles away, and many new sights to see, my mind was with my dear little sweetheart who was in St. Louis.*
>
> *Before this you have received my letter of Wednesday, and no doubt you laughed over it too. Perhaps it was foolish to write such a long letter with so little in it, but the fact is when writing to you it is a task to stop. I am sorry I have no news to write, but I have not been out since I returned, so I have not seen anyone you know. George was here yesterday afternoon so we went for a ride on our auto, we went out near your house, but everything looks so strange to me, and it seemed an age since I was out that way last. Yours devotedly, Jack*

What Jacob could not yet know was how important the internal combustion machine would be to hauling the product that would make him prosperous.

CHURCH LIFE

At that time work and love, together with social and spiritual life, tended to define a person's character. Jacob Best was an industrious young man looking for ways to maximize his entrepreneurial skills. He had fallen in love with a beautiful woman, and together they enjoyed such dating activities as dancing, boating and dinning in turn-of-the-century Chicago. Just as important was their spiritual life. Both Jacob and Hulda had attended St. Pauls German Lutheran Church in Lake View from an early age. Jacob in particular demonstrated a religious sensibility as this letter from August 19, 1903, shows. "But sweet one, I don't like you to miss church, remember you said if Miss Nofrinze wouldn't go you would go alone. I can recall Saturday nights that you were out quite late but you were never too tired to rise for Sunday school. Won't you go next Sunday, go just to please me, will you?"

From age nine, Jacob Best had attended St. Pauls. The church dated to 1843, when a group of residents secured a plot of land on the Near North Side from two wealthy Chicagoans, first-ever mayor William Ogden and his friend Walter Newberry. The two donated property at the southwest corner of LaSalle and Ohio Streets, and the development of St. Pauls Evangelical United Church was underway.

Church membership grew thanks to first-and second-generation Germans, eager to retain their Old World heritage. Congregations turned their churches into a kind of insurance policy. During much of the nineteenth century, many aspects of social welfare in

America were left to religious denominations. So it was only natural for St. Pauls in 1869 to open the German Orphan Home with a gift of land from Mr. and Mrs. Carl Uhlich. And St. Pauls was enough of a community fixture by 1871 that the building's destruction in the Great Fire was but a temporary setback. In 1885 the congregation welcomed Rudolph John, who could speak both German and English, as its new pastor. In 1896 the congregation moved to Lake View. Thirteen years later, Pastor John gave the first English sermon at St. Pauls. But it would not be for another twenty years that the congregation held English and German services on Sundays.

St. Pauls provided for both the spiritual and social needs of its members. Many church sponsored picnics were held in nearby Lincoln Park, which had developed into a model open space in urban America. Boating, swimming, concerts and other city events were held on weekends. Women with parasols were joined by men in their new boater hats for a stroll through the park. This, Jacob knew, was what the smart set did on a Saturday or Sunday-after church, of course-in Chicago.

Here is an invitation Jacob saved from St. Pauls for a picnic on Saturday, August 29, 1903. Revelers were to meet at the foot of Fullerton Avenue, within sight of Lake Michigan:

St. Pauls Church Guild
Orchard & Kemper

To Members And Friends:
You Are Cordially Invited To Attend A Lawn Fête Given By

St. Pauls Church Guild At Lincoln Park.
Saturday, August 29, 1903
At The Foot Of Fullerton Ave., From 4 P.M. To ______
In Case Of Rain Fête Will Be Postponed
Respectfully, Wm. R. Uhlmann, Chairman
Ladies Carry Lunch.

The April Open Session was yet another church-sponsored activity:

St. Pauls Church League Orchard & Kemper
APRIL OPEN SESSION.
You are cordially invited to attend the Open Session.
Thursday, April 21, 1904.
THE DRAMATIC CLUB
presents the bright comedy
Mrs. Jones' Butler

St. Pauls Orchestra (Mr. C. A. Weiss, Conductor) will render a bright program and Miss Carrie A. Letz will sing.

No admission fee will be asked, but it is thought, that the audience will gladly help to defray the necessary expenses by a contribution at the close of the play.

A pamphlet recently published by the League will be presented to all attending.

Charitable benefits were another part of Chicago social life. Hulda and Jacob often had the opportunity to attend an event such as the spring 1904 fundraiser by the Penny-A-Day Club for the German-American Hospital. Like the clubs, these events allowed ambitious people to make contacts in the world of Chicago business and politics.

But there was more to Hulda's life than helping her beau make the right contacts. These two letters from her parents give a sense of domestic life for a prosperous German family in 1903 Lake View:

WM. HAMMERSTROEM COAL: 98 WILLOW STREET, CHICAGO.

August 17, 1903

I have received your letter of Aug. 15, 1903 and I am glad to hear from you, that you and your friends all well, the reading of your letter has pleased me very much. If you are enjoying your outings which you have taken with your friends very much, I hope that you have had a good time yesterday on the Mason Picnic with all your friends which you have made there we miss you in St. Louis. Myself, Papa, Geo and grandpa are all well.

Hopping to hear from you soon, and best wishes to all your friends, Yours truly, Mama.

This letter from Hulda's father brings a little comic relief, showing how the big city could have a bit of the farm in it:

Your mama and Mrs. Nofrinze left this morning, Aug. 19th, 1903 to Aurora, IL to the Erholung's Picnic where they gone to have a good old time. Mama said to me that I have to stay at home today to take care of the house, and all the work she told me to do while she is gone to the picnic, first to take care of Beauty and Nixey and feed them well, 2nd feed the chickens and watch out for the Big dog in the ally that he won't catch the chickens and take good care of the horse, and clean the birds cage and feed them and scrub the front porch and then sprinkle the flowers and cut the grass and trim it well. Just think of it then she told me to prepare a good supper for her when she comes back from the picnic, it is one blessing that she did not tell me what to cook for her, so I had my own way to make a good supper for mama and myself, and George was not in it, and he missed it, so I start at 5 o'clock to make supper…When the fire department came rushing around in the kitchen door, they asked me what are you trying to do so I told them that I was just trying to tackle one of them big potatoes Parma cake, so they left again for their headquarters this is one day experience of housekeeping…Hoping that your happy and gay and to hear from you soon. Goodbye, Your Papa.

No one in Lake View wanted another Great Fire!

JACOB BEST AND AN UNCERTAIN FUTURE

Jacob Best was no different than other Chicagoans in wanting the kind of success that Horatio Alger wrote about so much. While it would be unfair to say Best married his future, it is clear that William Hammerstroem had important connections in the Chicago coal business.

Hulda wrote Jacob on September 5, 1900, with news that her father had been injured:

> *Father's hand is quite serious, yet as it had to be lanced and he is unable to attend to business and as it happens to be the right hand, it makes it very inconvenient to him and so I have to take his place in the entire business and my time is very much occupied. I have to be down to open and also close the office and my hours are quite long more so, as I am not accustomed to it. From 6:30 to 6 o'clock, and I think there are two more weeks before me. Well it won't do me any harm as I shall learn the entire business from beginning to end, and I enjoy it that I have all the managing to undertake. Your sincere friend, Hulda.*

Best responded three days later. His letter is a mix of sympathy and envy:

> *I am very sorry to hear that your father's hand is so serious, and that you have such long hours, you are certainly very sensible for thinking of it as you do, for I think there is nothing like a business education either for a lady or gentleman. I am almost discouraged with my position [at Marshall Field's]. Mr. Hunter is so different from what I thought he would be, that it makes it very unpleasant, and my work being very little just at present, the hours seem like days to me.*

Jacob Best and family visiting brother George in Chicago.

Four months later, in January 1901, in another of his courtship letters, Jacob found it impossible to hide his dissatisfaction with work.

> *Five weeks have passed and I have not seen or heard from you, and as I often think of you and wonder how you are, I take pleasure in writing these few lines trusting that you will kindly pardon me if I intrude. I suppose you are quite busy at the office as the weather is pretty cold but then I guess you make up for it going to masquerades and dances and have lots of fun. I have not been to a dance since I last saw you. Maryanne and I were at a Novelty Party last week, talk about fun. You might imagine the ladies all brought hats and trimming, and the gentlemen had to trim them for prizes. I got second prize, so don't say I can't trim a hat, but "O" such hats...Business is very quiet in our department at the store, but I am always busy getting in and out of trouble with Mr. Hunter. Yours sincerely, J. Best*

On August 19, 1903, Jacob wrote Hulda, "In reference to taking a position out of the city, I can't say. I can do better financially but I don't know what I will do, will tell you more when you get back." Risking his relationship by leaving town testifies to the growing desperation Jacob felt. Almost two decades later, he recalled that frustration. Working so long in the fur department at

Marshall Field's made him realize that "when labor is intelligent enough to avail itself of ideas, wealth may be created in considerable quantities. Until then, labor however conscientious and diligent, can do little more than produce the bare necessities of life."

Seventeen months after his wedding, Jacob finally left Marshall Field & Company to try his hand pushing a greengrocer's cart through his Lake View neighborhood. Chicagoans depended on the street peddlers for the convenience they offered in bad weather and on days when parenting made a trip to the store all but impossible. Jacob may have worked on consignment through a wholesaler on South Water Street, where all the meats, vegetables, fish and other food products were sold wholesale in open-air markets. The food came in at night and the peddlers by dawn.

Best's competition was the local neighborhood butchers and grocers who seemed to be located on every corner in the neighborhood. He was able over time to trade in his cart for a store. The establishment was decidedly pre-supermarket. Most households at the time looked for the following: a copy of the *Tribune*; pickles and tobacco for 5 cents; coffee for two weeks, 30 cents; oranges and cherries, 15 cents; gum, 3 cents; flour, 78 cents; candy, 25 cents; Cracker Jack, 5 cents; ice cream, 15 cents; and beer, 5 cents. If the grocery list is any proof, people had simpler tastes.

Jacob and Hulda lived at 2806 Southport, where they paid a monthly rent of $16 and a gas bill of $2.20. While not the lap of luxury, the Southport apartment gave Jacob time to think and to plan. As the child of immigrants, Best had loftier dreams than a corner grocery. He already had the necessary inspiration in the love of a good woman.

Jacob Best' first horse-drawn coal cart, 1909. Best soon learned that the coal business suffered from an image problem. In response, he marketed Best Coal as a state-of-the-art (and spotless) supplier of energy.

CHAPTER 3

THE BEST COAL IN CHICAGO, 1908-1963:
Jacob Best as Businessman, Philanthropist and Civic Leader

JACOB BEST: A SHOVEL, A HORSE AND A CART

Jacob Best did not simply want to earn a living in Chicago. He knew the great Marshall Field as an employer and no doubt met him as an employee. And the mere mention of names like Armour, Swift and Pullman served as motivation. Jacob worked to succeed just as these famous Chicagoans had. But, he was barely surviving from paycheck to paycheck. It was time to find another way to leave his mark. There was something about coal.

Historian Donald Miller explains the importance of this now largely ignored energy source, "America's, and Chicago's, industrial economy was a creation of incredible diversity, making everything from straw hats and pianos to bridges and boilers. The key to it all was coal. Coal is a city-forming material. The coal supplies of Southern Illinois, first discovered in the 1850s by geologists of the Illinois Central Railroad, made possible Chicago's fantastic industrial growth between 1860 and 1890, when the number of workers employed in the city's industries rose at least thirty fold. Cheap, accessible fuel, the first requirement for mastery in manufacturing, allowed Chicago to maintain its transportation advantages over both water and land. Coal induced scores of industries to locate in Chicago, including the railroads that ran on coal. Coal attracted two types of industries to Chicago: those, like the iron industry, that used coal for heating purposes, chief among them brewing, canning, and baking, and those that used it to feed steam engines that ran rows of fast-moving machinery, such as plants for making shoes, ready-made clothing, and printed material of all kinds."

In 1909 Best joined in a partnership with his father-in-law William Hammerstroem, a longtime veteran of the coal business. Best was the de facto apprentice working to learn the business. There was one more thing. This apprentice yearned to be a master at his trade, ASAP.

Nearly a half-century later, author Muriel Herzberg detailed the beginnings of the Best Coal Company. "Jacob Best, at the age of 30, walked into the office of a North Side businessman and asked him for tips on how to become his competitor. Best had been a law office clerk, lithographer's apprentice, furrier's assistant and grocer before his uninvited request to Francis S. Peabody, then president of Peabody Coal Company. Peabody not only gave Best ideas about the coal business but gave him some funds to get a retail coal firm started. Within 12 months and the sale of almost 10,000 tons of coal, Best was able to repay Peabody…The 10,000 tons of coal sold during the first year Best was in business were chiefly anthracite, a hard coal used for home heating; Pocahontas, used to heat apartment building flats, and Southern Illinois nut, used for stove heat. It sold for $3 to $4 a ton and was delivered in large chunks carried in horse-drawn wagons."

In all the Horatio Alger stories, the protagonist takes advantage of some bit of serendipity. For Jacob, Francis Peabody proved to be the luckiest break in a lifetime.

The Edgewater Yard, birthplace of Best Coal. The Mayfair Yard on the Northwest Side would be nearly twice as large.

Over the years, wagons gave way to trucks at the two Best Company coal yards. The facility at Bryn Mawr and Broadway was typical. Best had used the loan from Peabody to get his start and then from the proceeds of sales he built an elaborate coal yard. Company literature from 1929 noted how the coal was "delivered at the yard over the North Shore line of the Chicago Milwaukee & St. Paul railroad, and was spotted on two switch tracks, several hundred feet long and sixty feet apart, paralleling the railroad. This enabled the company to handle ten cars at a time with ease. These switches were overhead, branching off from the North Chicago elevated railway line, and the space underneath was given over to fifteen pockets necessary to unload cars of hard coal. All a driver had to do in loading was to place his truck under the opening to the pocket, pull a lever and get a load of clean coal. With this method it took five minutes to load eight tons."

The Black Diamond writer visited the Broadway yard to report, "I was impressed at once with the cleanliness of the place. Not a speck of dirt. A clean mosaic tile floor. Shining showcase and counter. Orderly rows of polished and varnished desks. Counters of black Georgian marble: the front of the office well lighted and wide comfortable, substantial seating provided for customers." The writer admitted there was "nothing ornate about the office, but an expensive interior decorator would probably be lost to find suggestions for improvement, considering the nature of the business." Coal companies, needless to say, tended to care very much about image.

The office of Jacob Best Coal at the Edgewater Yard. A candlestick phone sits on the front desk.

GERMANS IN CHICAGO

Jacob brought but one weakness to the business-his ethnicity. It was that or his timing. Germans were the second largest immigrant group in Chicago at the opening of the Civil War. By the turn of the century, they jumped ahead of the Irish to rank as the city's largest ethnic group. With size came confidence. Germans prided themselves both for their ability to assimilate into American society and yet maintain their Old World identity. But with the outbreak of war in 1914, an increasing number of Chicagoans began to think there was too much overseas attachment.

On April 6, 1917, the United States declared war on Imperial Germany. In a sense, the declaration carried over to Lake View. The environment that Jacob and Hulda had grown up in was suddenly suspect. German clubs and German culture were now considered alien, even treasonous. Anti-German protesters vandalized a statue of Goethe, the great German thinker, in Lincoln Park. The hostility did not let up until the signing of the Armistice. In the words of historian Melvin Holli, "Like a violent act of nature, a broad, unbroken, popular front of newcomers and native Americans united to bulldoze and expunge a perceived evil from their cultural landscape."

For years, people and things German bore an unmistakable stigma. The Chicago public schools dropped German instruction for elementary grades along with a school speller featuring a lesson on the virtuous future Kaiser Wilhelm. As went the speller, so went Bismarck School, with the new name of Funston. Emotions ran just as high outside of school, with Chicago Symphony Orchestra conductor (and German immigrant) Frederick Stock taking a leave of absence in order to finish his naturalization. Stock was just one in the mass of German Chicagoans expected to behave accordingly, which was patriotically, at home, work and play.

Suspicion fell on Jacob and his family if only because of their house of worship. At St. Pauls the sermons were still delivered in German-with representatives of the Federal Bureau of Investigation on hand to monitor them. It was during this period that the church decided to go by its English, rather than German, name. As a church history notes, the people of St. Pauls were among the 400,000 German-Americans in Chicago who had to reconcile mixed feelings over "American patriotism, pride in German heritage, and the Christian will for peace" stemming from the German war effort in Europe. When the United States declared war on Germany in April 1917, the congregation "rallied behind the Stars and Stripes and Americanism was the theme for an evening 'full of patriotism and historical suggestion,' sponsored by the women of St. Pauls Church League."

That Jacob Best's new coal business survived at all was something of a miracle.

POST-WAR CHICAGO AND THE COAL INDUSTRY

Ethnic prejudice was one problem for Jacob. Pollution was another. There was no getting around the fact that coal made a mess. Burning soft coal generated smoke while burning any kind of coal left behind cinders that had to be cleaned out of furnaces and disposed of. Even under the best of circumstances, Jacob found himself part of a business held responsible for creating the "Smoke Evil" that hung over so many American cities.

Coal "hiker" loading coal from shaker # 1 in 1924.
A coal yard provided vigorous employment, to say the least.

In this business environment, success went to those who offered the best product with the best service at the cheapest price. Jacob increased his coal distribution seven-fold during the early 1920s, so he had to be following this formula. He tried to address the pollution issue by advocating the installation of automatic coal stokers in furnaces as a way to increase energy efficiency while reducing smoke levels. But coal falling off a delivery also created a nuisance. So did the feces of horses pulling those wagons. Jacob found his solution by replacing horse power with truck horsepower.

A side view of shaker # 1 in the Mayfair Yard.

The kelly-green delivery trucks were yet another innovation by Best Coal. Best grew his business to where he could afford sixteen trucks and crews. Each truck had a capacity of from five to fourteen tons of coal. Deliveries ranged on the south from Division Street north to Evanston, and from the lakefront west to Harlem Avenue.

The industry journal *The Black Diamond* chronicled the company's growth, "With new apartments and business buildings as well as scattered manufacturing plants between 1909 and 1924, the Edgewater coal merchant increased his tonnage to beyond the 70,000 mark the latter year, emerging as one of the city's leading retailers." Best was now able to approach prominent downtown bankers to expand his business enterprise. He established relationships with Harry Wheeler of the Union Trust Company and John W. O'Leary of the Chicago Trust Company.

In 1923 Best opened a second yard at Kilpatrick and Berteau Avenues on the Northwest Side. The new Mayfair facility provided in excess of 100,000 square feet of space compared to 60,000 square feet at the Edgewater Yard. Best employed a workforce of nearly fifty drivers, helpers, yardmen and hikers on a typical winter day.

THE YEARS 1925-1941

On August 4, 1927, *The Howard*, the official newspaper of the Howard Avenue District Business Association, reported that, "In conjunction with the celebration of the fifth anniversary of Howard Avenue Trust and Savings Bank, bank officials announce the election of five prominent local men as directors, thus increasing the board to 15 members. The financial strength of the institution may be estimated by the bank's achievement of having established a record of over $1,470,944.59 in assets as of June 30, 1927. New directors are men successful in commercial fields and civic life. They [include] Jacob Best, president Jacob Best Coal Co., past president Rogers Park Kiwanis Club a prominent businessman in

Rogers Park [known] for his business success and interest in civic work." With the announcement of his bank appointment Jacob Best had arrived on the Chicago business scene. Jacob celebrated by taking a family vacation. Accompanying him and his wife on a three-month tour of Europe were son Thilo, born November 17, 1906; their second son, also Jacob, born September 15, 1909; and a daughter, Bernice, born July 31, 1914.

Jacob Best' home at 6338 N. Kenmore in Rogers Park, Chicago.

Jacob's and Hulda's Lutheran faith played a significant role in raising three children. A family heirloom gives evidence of the kind of home the Bests kept. It is a copy of the *King James Bible* given to Hulda's brother George by his Sunday school teacher, Miss

Stoetzel, on November 1, 1891. There was an inscription on the front page entitled *"Home"*:

It takes a heap o' livin' in a house t' make it home,
A heap o' sun an' shadder,
an' ye sometimes hace t' roam.
Afore ye really 'preciate the things ye lef' behind.
An' hunger fer'em somehow, with 'em allus on yer mind.
It don't make any differunce how rich ye get t' be,
How much yer chairs an' tables cost,
how great yer luxury; It ain't home
t' ye, though it be the palace of a king,
Until somehow yer soul is sort o' wrapped 'round everything.

Ye've got't sing an' dance fer years,
ye've got t' romp an' play,
An' learn t' love the things ye have by usin' 'em each day;
Even the roses 'round the porch must blossom year by year.
Afore they 'come a part o' ye,
suggestin' someone dear,
who used t' love 'em long ago,
an' trained 'em jes't' run.
The way they do,
so's they would get the early mornin' sun;
Ye've got't love each brick an' stone from cellar up t' dome.
It takes a heap o' livin' in a house' t make it home.
By Edgar A. Guest

Jacob Best and his daughter Bernice, 1916.

Jacob and Hulda with children (from left) Thilo, Bernice and Jake.

By the same token, it took a lot of coal to make a house warm enough to be a home. In 1906 *The Daily News Almanac* reported that 63,698,803 tons of anthracite and 306,084,481 tons of bituminous coal were produced in the United States. By 1927 those figures had increased to 72,011,000 tons of anthracite and 519,804,000 tons of bituminous coal in addition to 50,925,000 tons of coke. The Daily News put the overall value of coal production at over $1,471,963,000. Coal grew Chicago, and Jacob Best sold coal.

Best took an important lesson from his time at Marshall Field's, that a man was only as good as his word. It was reported that, "Best on one occasion took a $5 a ton loss on thousands of tons of coal in order to fulfill contracts he had made with his customers. If called upon today [1929] to do the same thing again, Jacob Best wouldn't hesitate in fulfilling his obligations. Impartial dealing with customers and integrity carried him a long way in the coal business, and he was convinced that it is far too late to abandon these virtues now." During a 1923 coal strike, coal was sold for $12.42 per ton and Best had many orders on the books for $7.50 per ton. Best honored his contracts and delivered the coal at a substantial loss. He not only honored his word, but he sold all his coal with a money-back guarantee if customers were dissatisfied with the quality of the fuel or its delivery.

Jacob Best truck fleet at shaker # 1 in the Mayfair Yard. Trucks conveyed a sense of modernity to customers-and they were cleaner than horses.

Best's competitors thought enough of the "rags-to-riches" second-generation American to elect him president of the Chicago Coal Merchants Association, the largest association of its kind in America. He also served as a director of the National Retail Coal Merchants Association and chaired that group's 1929 convention committee. Assisting the founder of Best Coal in the business were his sons Thilo at the Edgewater Yard and Jake at the Mayfair Yard. The business was very much a family affair.

Jacob 'Jake' Best, son of Jacob Best coal magnet, was manager of the Mayfair Yard.

On September 13, 1927, this advertisement crafted by Best was sent to his customers. It shows Jacob's business acumen:

Dear Friend:

All of us are interested in the making and saving of money. There are many ways that this is done but there is no more satisfying way of saving money than buying the necessities of life in such a way as to derive the greatest benefit for the amount of money spent. A dollar saved in 10% interest on a ten-dollar bill and you save many dollars on your fuel cost when you buy your fuel from Jacob Best Coal Company and take advantage of the service that Best gives with every order of fuel.

Ten per cent is good business when it is made in a year, but 10% made in six months is still better business. Why not make your fuel bin pay you money? We offer you that opportunity.

We especially recommend that you fill your fuel bin with Genuine Koppers CHICAGO Coke. This excellent fuel has all the advantages of the best Anthracite and not one of the disadvantages, costs you less money ton for ton and gives more heat. It is easy to use and control and easy to learn how to use.

Jacob Best Coal Company maintains and operates a Service Department of trained heating men. These men will be glad to discuss your heating problems with you free of cost and they will be glad to explain coke to you. Just phone us when you can see one of these men.

Yours truly, JACOB BEST COAL COMPANY

In 1929 *The Black Diamond* reported, "During the nineteen years since it was established (1909) the progress of this company had been constant. When it was first organized all coal it marketed had to be shoveled, screened and loaded by hand." Anyone wanting to meet this success story in person had the chance to do so at the twelfth annual convention at the Stevens Hotel in Chicago in the spring before the Crash on Wall Street.

On May 25, 1929, *The Black Diamond* reported, "Next Monday morning at 9:00 o'clock daylight saving time, the starting gong will ring on what promises to be the greatest and most enthusiastic gathering of retail coal men in the history of the trade. Gathered under the banner of the National Retail Coal Merchants' Association, hundreds of merchants from every corner of the national map will be in attendance. And, thanks to the energy of the several national and local committees, the convention will be ready, Jacob Best, president of the Chicago Coal Merchants Association and general chairman of the convention committee, and James D. Rooney, executive vice president of the Chicago association and vice-chairman of the convention committee, have left no job unfinished."

Jacob Best, Coal Merchants Association President.
(Chicago Tribune photo)

Like other business conventions, this one had all sorts of entertainment and sight-seeing opportunities. It also had a serious side, as evidenced by one of the addresses, on *"The Coal Industry Re-Discovers Automatic Heat Control."* Already by 1929, there was a growing sense that coal was a business in danger. *The Black Diamond* reported that:

> *If we are to compete with the other industries that are storming our gates we must give the public what those industries offer: convenience and cleanliness. The old idea of dumping coal in a basement and considering the job finished is gone forever. The men who are going to talk to you today, tomorrow, and Wednesday, are going to touch on different phases of our new job in serving the public.*

One participant declared that, "a campaign such as the building trades are fostering to bring the sixteen million antiquated American homes up to date would do much to revive the coal industry's prosperity by increasing the total consumption of coal…the coal retailer should sell heat and instruct the homeowner how to use it. Instead of assuming that his responsibility ceases when the coal is dumped in the bin, the coal merchant should feel that his opportunity for merchandizing has just started. Care should be used in selling the proper size and quality of fuel for the heating plant use by the consumer."

The coal business pinned its hopes for the future on coke, a hard burning fuel long used by the steel industry that had become available for residential consumption. The basic process for producing coke starts with a blend of metallurgical quality coal. The fuel is pulverized into a powder of particles smaller than grains of sand. Tons of coal powder are then dumped into coke ovens. The ovens are made of brick, and, when the powder is placed inside, the lids are sealed so a temperature of more than two thousand degrees can be attained. The typical baking time is eighteen hours, and this intense baking turns the coal into coke, a porous material of almost pure carbon that burns much more intensely than regular coal.

Coal "hikers" hard at work in the Edgewater Yard.

Best and the rest of the industry hoped that coke would allow retailers to generate an extraordinary demand among twenty million coal users. Other retailers hailed Jacob for his "rare executive ability coupled with a broad understanding of the problems of the retail fuel trade." These would be words to draw strength from once the Great Depression hit.

Along with a convention to run in 1929, Jacob had a twenty-fifth wedding anniversary to celebrate. On June 10, 1929, Pastor Rudolph A. John wrote Jacob from St. Pauls House, a Churchly Home for the Aged:

> *My dear friend:*
>
> *I would not be true to myself did I not send you an expression of the thoughts that were felt in my heart at the really splendid and altogether unique celebration of your Silver Anniversary. I am sure that none of your guests can ever forget that day. Usually it is expected that the friends bring silver tokens as gifts to such a feast. You have completely reversed the established order of things. You gave to your guests not only a delicious dinner but also that very beautiful entertainment at the theater. When the final curtains fell the old man declared: "I have had a wonderful birthday," I wondered whether you did not know that wonderful thought in your heart when you said "I have had a wonderful Silver Wedding Day!" It was very very beautiful and I deem it a privilege to have*

been with you. I thank you most cordially, for the very generous gift that your envelope contained. I could not accept it as a fee for professional services, but only as a heartfelt expression of a friendship which means more to me than I can express in words. You know what I mean and will understand fully. I am so thoroughly glad to know that far away was indeed a way of happiness. Splendid physical and mental health, children of whom you can justly be proud - you and your wonderful wife must be very very happy - and that makes me happy, the old friend who has shared so many years of sunshine and love with you. I can but repeat, "may your silver be changed into gold." I will not be there then but I trust to God somebody will be there to speak of the day in 1929 when your eyes shine and your heart beat high at the silver wedding

Gratefully and faithfully, I remain your old friend

Rudolph A. John, Pastor

Jacob did more than treat a friend from St. Pauls. He also helped raise $200,000 needed for the construction of a new Uhlich Home for Orphans. His own experience as a child at Angel Guardian Orphanage made Jacob sensitive to orphan needs. In addition, he agreed to serve on the board.

Best soon celebrated yet another Silver Anniversary, this one

Hulda Best at her home on Kenmore Avenue in Chicago's Rogers Park: Loving and lovely.

with the Chicago Coal Merchants Association. On February 16, 1931, he was honored for his many contributions to the organization. Jacob had more than enough reasons to give thanks, both in family and business.

The Depression meant any number of challenges for the business. This May 12, 1933, advertisement for Best Coal was an attempt to increase volume through an "early bird" sale:

> *For more than twenty-five years, Jacob Best Coal Company has been helping its customers heat their homes satisfactorily, at low cost.*

Mayfair Yard shaker # 2 during a Chicago winter. The worse the weather, the better for business.

Today, in line with this established policy, we announce LOWER PRICES on all our high quality fuels, including genuine Koppers CHICAGO and Solvay CHICAGO coke.

In our opinion, present conditions will force basic commodity prices upward, and we therefore advise you to purchase your next winter's fuel requirements at today's low prices.

Telephone Sheldrake 2060, have Betty Davis of our personal service department arrange for an inspection of your heating plant without cost by one of our service men.

Confer with him regarding the best kind and size fuel for greatest economy, and take full advantage now of the Plus Value we offer.

Sincerely yours,

JACOB BEST COAL COMPANY

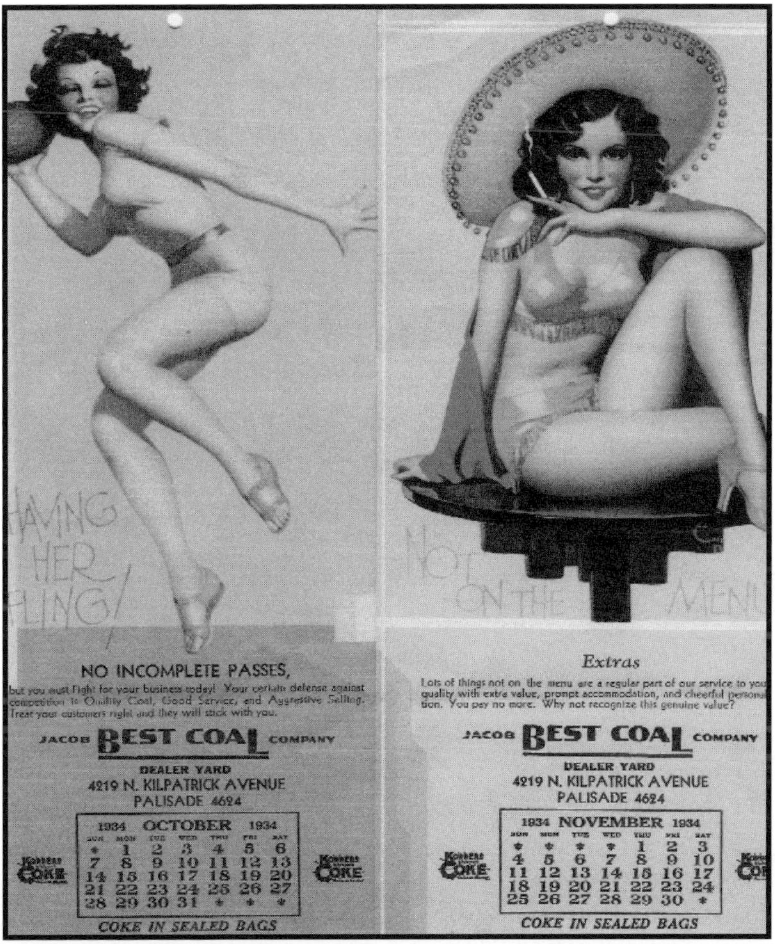

Jacob Best Coal Company advertisement calendar. Long before sex sold cars and soft drinks, it had a go-around with coal.

Best also wrote to his customers that:

> *Every member of our organization joins with me to express our appreciation and to thank you for the orders that you have placed with us during the past year. For a quarter of a century our company has been serving thousands of satisfied customers. We are striving every moment to render better service if possible and to merit your continued confidence and favor. Labor difficulties have been adjusted, coal prices and labor charges are materially less than they were last year. We are therefore looking forward with courage and confidence to make your association with us more profitable and satisfactory. We have every reason to believe that early buying this year will prove exceptionally profitable, and respectfully urge your consideration at this time.*
>
> *Sales tax paid by our customers will be refunded at the earliest possible moment. You are paying for the BEST COAL and it is our policy to make sure you absolute satisfaction. It is very seldom that we receive a complaint of any kind. If, however, at any time, now or in the future, you are not completely satisfied with Best Coal and Best Service please tell us. We will do everything in our power to make every detail right. It is upon this policy that we have built our business.*

The above ad is interesting as much for what it implies as says. Like other businessmen, Best was trying to turn the deflationary price spiral of the early Depression to his advantage. But he did more than take credit for low prices. He gambled that first-rate customer service would keep him in business, and he won.

Advertising on wheels, 1938.

The difference between success and bankruptcy in the 1930s could hinge on something as simple as the lettering on two new trucks. *The Black Diamond* reported in 1938 that, "Jacob Best, president of the Jacob Best Coal Company, one of Chicago's outstanding coal merchants, has found a new way to attract the favorable attention of his customers and prospects. Recently it became necessary to add two trucks to the firm's fleet…The words 'BEST COAL' are built of chromium nickel, and the word 'always' is a brilliant red. The combination of raised bright silver and red letters against the green background invariably attracts the attention not only of customers when the truck is making deliveries, but of passers-by who frequently crowd around so thickly that the scene resembles a political rally. Mr. Best states that he has found it a sound business policy to keep his equipment in topnotch condition, and any faulty equipment is promptly discarded."

Despite the uncertain economy, Jacob did not ignore the responsibilities of his faith. Rudolph A. John, former pastor of St. Pauls and now chaplain of St. Pauls House, praised his friend as a man trying to do good in this letter dated December 10, 1935:

> *My Dear Friend:*
>
> *Your generosity has overwhelmed me and I hardly know how to find words to express my gratitude. The large check which you have sent me was a personal Christmas gift to the evidence of our friendship which I value beyond words and can never forget. There has never in all my life been an incident which touches me so deeply. In my helplessness and suffering it has burst upon me like the light of a great star shining in the darkness. I thank you with all my heart and can but hope and pray that a good God will reward you and grant you a full measure of happiness as long as you live. I will not be here much longer and when I am gone I hope you will continue to think of me as one who has always admired and loved you.*

With every Christmas wish for you and your family, I remain gratefully your old friend.
Rudolph A. John, Chaplain

Best also provided financial support to the Ulhich Orphanage at 3730 N. California.

ANOTHER WAR, 1941-1945

World War II began for the United States on December 7, 1941, with the attack on Pearl Harbor. On December 11, America declared war on Japan and Germany. If at first some German Americans were divided in their feelings about the old country, which had long since changed by the time of Hitler's suicide in April, 1945. Their world was transformed, which is to say the world of Lake View and the rest of German Chicago.

For two generations, the Best family lived in a city that, while of the New World, maintained many elements of the Old. Language, institutions and culture all continued as if Lake View and Lincoln Avenue were parts of Berlin. World War I ended this environment that was so easily German and American. It revived, to an extent, during the interwar period only to be crushed by the horrors perpetrated by Hitler. In essence, just one Nazi apologist was enough to ensure that German ethnicity after December 7, 1941, would in no way be like that encountered by the first Jacob Best when he came to Chicago in 1869.

Unfortunately, in 1936 St. Pauls' pastor received the

"Distinguished Service Cross of the Medal of Honor of the German Red Cross from Fuhrer Hitler and the Third Reich, for his support of the fatherland, and his work among German/Americans." In the wake of Pearl Harbor, such an award was both unthinkable and humiliating. The need to prove one's loyalty may explain why 205 men from St. Pauls fought against the Third Reich and twelve of them made the supreme sacrifice.

While the war raged on in Europe Best provided coal on the home front and began his philanthropic career. To his credit, Jacob Best was able to make his way in this changed world, to the extent

Best coal being loaded during World War II, Jacob Best doing his part for the War effort on the home front.

that he was appointed to the board of the Chicago Public Library on June 23, 1943. If Chicago was an intolerant place in wartime, it apparently stopped at the book stacks. Mayor Ed Kelly wrote Best:

> *OFFICE OF THE MAYOR*
> *CITY OF CHICAGO*
> *Edward J. Kelly*
> *Mayor*
> *Mr. Jacob Best*
> *6338 Kenmore Avenue*
> *Chicago, IL*
> *Dear Mr. Best:*
> *It is a pleasure to inform you officially that the City Council has today confirmed my appointment of yourself to membership on the Chicago Public Library Board.*
>
> *You are rendering a genuine public service in accepting this important responsibility, and I am confident that your membership on the library board will prove most beneficial to that organization and to the citizens of Chicago.*
>
> *With cordial good wishes, I remain*
> *Sincerely yours,*
> *Edward Kelly*
> *Mayor*

In doing public service, Jacob showed his loyalty to city and country in the most difficult of times.

JACOB'S AND HULDA'S FAMILY

Jacob Best with Hulda and friends out bird watching.

Best focused his life on business and family during the turbulent war years. At Sunday dinner, the family gathered around the table to pray, "bless the food to the strength of our mind, body and spirit," before eating. Family traditions, not least of all the holidays, sustained Jacob, Hulda and the children when they most needed it. But not even the love of family could prevent tragedy.

In early April, 1945, Hulda went to Presbyterian St. Luke's Hospital for treatment of an anemic condition. One morning Jacob left his office and drove to the hospital to have lunch with his wife of forty years. He then returned to the coal yard, only to get a phone call. It was the hospital calling to say that Hulda had died after receiving a blood transfusion that did not match her blood type.

A friend (from left) with Bernice, Hulda, Thilo and Jake.

Jacob took to writing his daughter Bernice in his grief. A series of letters passed between them for three months, starting on June 27, 1945. Jacob wrote Bernice at Knoxwood in Macatawa, Michigan:

> *Dear Bernice:*
>
> *Uncle Max and I had a very lovely drive back made much better time but were indeed happy to get back to Chicago. I neglected to tell you that I received my father's day gift, the beautiful address book which was indeed a long felt want which I appreciate very much and for which I thank you.*
>
> *I am sending a letter to Ruth at this same writing. I did make one blunder I neglected to take the price tags off of the articles I bought. I know this is*

a terrible neglect but it is too late now to change and I hope Ruth will not feel uncomfortable in any way.

I would feel that you are not so very far away if you would drop me a card periodically. I really don't get a chance to get lonesome but I think of you very often. The Kubels were at the house Sunday before I got home and left a note for me and in the party were Richard, his wife, Richard's mother and Emma and Bill. I was sorry to have missed them but Bill and Emma came over for a short time last evening.

Uncle Max is very anxious for me to go to the church initiation tonight of Pastor Lotics. I am not keen about it but I guess I will go as it no doubt will please him very much.

The confidential matter that we discussed and over which I was so much perturbed has finally been disposed of but I finally had to get quite excited before this conclusion was reached. Let me hear from you when you can in the meantime I am always thinking about you.

Lovingly,

Dad

Neglected to mention that I enclosed Marsh & McLennan confirmation of liability coverage.

These are the words of a lonely man trying to find comfort in the details of everyday life. On July 11, 1945, Jacob wrote again:

> *Dear Bernice:*
>
> *You no doubt often wonder why I don't write to you. However, the fact remains that we go along from day to day in pretty much a regular routine. I think of you always and get your letters and postals that I appreciate very much and assure me of the splendid success that you and Ruth are having with the children.*
>
> *By this time you no doubt know that Jake has finally decided that Rhinelander should be his summer spot. Accordingly all plans were set for his taking possession on the 16th. However it now develops that matters cannot be cleared by that time and when he will get possession is doubtful. However when that is taken care of I have my blue jeans, rubber boots and fishing tackle all ready and looking forward to quite an interesting time with Jake and Tiny and his family.*
>
> *There isn't anything really new except that I might mention the fact that I finally became quite upset with Mollie and decided we should call it quits but she wouldn't be fired. Nevertheless we now have a better understanding and are getting*

along very beautifully. The shopping is still a big problem but I really believe she is making considerable headway because within the last ten days we have had steak, winnies, pork chops and lamb chops which, by way of comparison of what I hear, is doing pretty well. They hung the mirror in the living room yesterday and it looks swell and I am sure you will be pleased with it.

Since I have seen you I have had a very lovely note from Betty and no doubt you know by this time that the two boys were at home and there was quite some rejoicing at the reunion. Incidentally, on two occasions I have sent Betty some turtles and I have reason to believe they enjoyed them very much. While talking turtles, frankly do I keep you supplied or is there anything further or any courtesy that I might follow through with?

I make it a point to have Uncle Max for breakfast and dinner periodically and it more or less changes the monotony. In the meantime, I have been with the Rooses, Kubels, Halls, Hammerstroems and of course the family from time to time which is always a great satisfaction to me.

It might surprise you to learn that I have arranged with Mollie that she should have two weeks vacation and now when I thought we were all set it develops that she does not want to go away knowing how I am going to get along. The delay in Jake's transaction sort of upset our plans immediately but that of course will finally solve itself.

I see Jerry's letters and Barbara Ann's letters which are really remarkable for little kids. They are always interesting and confirm the fact that they are enjoying their Knoxwood vacation very much. I sometimes wonder what your plans are when you finally get set. I hope you will let me hear from you.

It goes without saying that when I think of Knoxwood, you and the children, I am also thinking about Ruth and finally I am inclined to feel with the responsibility that you people undertook was beyond reason and it will be interesting to hear just what your reaction of the entire episode is.

I might ramble on further but I guess anything I might say would be a repetition so with love and best wishes to you and Ruth,

Sincerely, Dad

A life of routine without Hulda must have been very difficult for Jacob. He had a big house which was once filled with the love of his wife and family. Now it was empty except for his housekeeper

Jacob Best and a day's catch.
The businessman was also a sportsman.

Mollie. Jacob had four grandchildren, Jerry and Scott from the marriage of Jake and Tiny (Clara Douglas Cornell) along with Thilo Howard and Barbara Ann from the marriage of Thilo and Barbara (Campbell). The grandchildren became the light of his life, and he cherished them and wrote them often. Jake and Tiny had a summer home in Rhinelander, Wisconsin on a lake with great walleye and pike fishing. Jacob was an avid outdoorsman, and he loved to fish the lake and visit with his family. Life now consisted of his children, grandchildren, the Coal Company and St. Pauls Church. On July 25, 1945, Jacob wrote Bernice:

Dear Bernice:

It was indeed a pleasure to get your very lovely letter. I left here with Jake and Tiny a week ago Monday and got back to Chicago Sunday evening. I had a very happy time and Jake has a very beautiful spot although it is very crude and it will require a lot of effort and expense to make it comfortable, yet both Jake and Tiny and the children seem to enjoy it tremendously.

Concerning you and Ruth coming in on Monday, that would be fine. I have no definite plans. Mollie has been on a vacation for two weeks. She called up yesterday and it seems she would rather come back as of today or tomorrow and take advantage of the full two weeks which I had promised her. I am getting along very nicely with Mollie excepting however, as you know, she lacks the managerial ability which is so necessary in making a home comfortable. However, there is material improvement and I am still hopeful that she will work out all right.

As for my coming to Macatawa, the fact remains I am most comfortably happy at home. I enjoyed the brief stay that I had there very much but immediately I do not feel I want to make the trip. The pictures which you enclosed are really pretty good, especially so of Uncle Max and Ruth.

I am awfully pleased that you sent them as they will really be souvenirs of our visit.

Barbara and Jerry surely were very happy with their stay at Macatawa. I did not have much chance to visit with Barbara Ann, but Jerry is everlastingly talking about it. I was amazed at the way Jerry could swim. Out at Rhinelander he was in the water every possible minute in fact he carried on like a regular water rat.

I happened to be at Aunt Tillie's this morning. They were very happy to tell me about receiving your card and I don't know whether you know it or not but Kathryn came in Sunday with the baby boy and will stay until the 6th or 7th of August. The youngster is surely a cute little lad and looks just like his father. Kathryn has not changed and is very happy to be with her mother.

It has been terribly hot here. However, we have all been waiting for it and are not much discouraged. Everything else is going along fairly well but I am sorry I have to confess that even now we have not got our curtains up. They are in the making however and maybe will be up by Monday.

With love and affection to you and Ruth,
Lovingly, Dad

Eventually Jacob's sorrow overwhelmed him, and he had to sell the Kenmore home because it reminded him of his lost love. One of the bidders on the house was Chicago Bears football coach George Halas, who lost the bid to another man who paid cash. Best moved to an apartment owned by his company on Michigan Avenue in suburban Evanston. He found solace in St. Pauls and his philanthropic interests.

St. Pauls German Lutheran Church of Chicago, the center of Jacob's faith life.

1950-1964: BEST AS CIVIC FIGURE

Jacob continued his philanthropic work and service on the Chicago Public Library Board. On June 15, 1950, Mayor Martin H. Kennelly reappointed him:

OFFICE OF THE MAYOR
CITY OF CHICAGO
Martin H. Kennelly
Mayor
Mr. Jacob Best
2071 Lunt Avenue
Chicago, IL

Dear Mr. Best:
I am pleased to inform you officially that the City Council has confirmed my re-appointment of yourself to membership on the Public Library Board.

You are rendering a genuinely fine and important service to the City as a member of this Board, and I appreciate your willingness to continue in this capacity.
With best wishes,
Sincerely,
Martin H. Kennelly, Mayor

And with Chicago civics came Chicago politics. Martin Kennelly was a businessman turned reform mayor. As such, he was a politician Jacob Best could understand. Jacob gladly contributed to the mayor's 1951 re-election campaign. The mayor won a second term, and Chicago proved ready for at least a little reform.

The eleven years Jacob spent on the Board of Managers of the Ulhich Boys Home was not quite as political. On July 19, 1952,

Pastor Grauer of St. Pauls Church, wrote to tell Best, who was stepping down, how much his work was appreciated:

> *Dear Mr. Best:*
>
> *At the Session meeting the other evening, it was reported that you are no longer on the Board of Managers, but had asked to be relieved of that responsibility.*
>
> *I would be remiss as minister of the church and especially as a friend of yours, not to take a moment to express my deep appreciation to you for your splendid services. I can remember eleven years ago, when I came, that you and I had some rather difficult times out at Uhlich and even though all problems have not been solved, and there are still certain things that you and I would like to correct, you made a tremendous contribution and surely you know that Uhlich is a different institution than it was ten years ago. You had more to do with that than you are ready to admit. I would like to thank you very much for the marvelous service you rendered the Home, and indirectly, the Church by your years of service at Uhlich.*
>
> *I miss the contacts with you since you were on the Board of Managers and I continued on the Board of Trustees and wish that some time or other when you are in town, that you and Lee Rockwell*

Jacob Best (upper far right) and fellow trustees of Uhlich Orphanage. Best also established a foundation to underwrite children's causes.

and I could lunch together. I talked to Lee about this the other day and we both look forward to such a little confab.

Hope you are doing well - that you are feeling fine. I am always glad to see you when you are in church although I often don't get a chance to speak with you. If there is ever anything that I can do, in any way, I know you will feel free to call me. Please accept my sincere appreciation.

Very sincerely,

Gerhard Grauer

Best's two sons, Thilo and Jake, were now responsible for running the firm, which had been renamed Jacob Best Coal & Oil Company; everyone in the business could see coal would soon go into severe decline as an energy source, to be replaced at least in part by heating oil. In 1954 Best was seventy-five years old, yet he still came into the office every day. It was said of him that he was as sprightly and vigorous as he had been three decades ago when the Best Coal Company recorded its first 100,000 tons in annual sales.

Mr. Lee McCann, Jacob Best and son Thilo Best at Edgewater Yard.

Best was now getting up in years. He had been witness to a great many events in Chicago from the 1893 World's Fair to the introduction of the motor car to the appearance of television. He had accomplished much and become an influential figure in the city he loved. On August 16, 1954, a good friend from the coal business wrote him to say:

> *Dear Jake:*
>
> *Miss McInerney who, as you know, has looked after my affairs for a good many years and is in this office, called my attention to the article in The Black Diamond recently about your company and yourself.*
>
> *I was very much interested in it and congratulate you on your success in life. There were some things about your early life that it did not tell but in the main it was a very good story.*
>
> *Why don't you call me up sometime so that we can lunch together once in awhile.*
>
> *Sincerely,*
>
> *C. Moderwell*

Best Coal truck with bagged coal for delivery.

Three weeks later, Peter I. Bukowski, president of The Cosmopolitan National Bank of Chicago, wrote Best:

Dear Jack:

The August issue of The Black Diamond, containing the article on Best Coal Company living up to its name for 45 years of its business existence, is a welcome reminder of the friendship that has characterized our relationship for these many years.

Congratulations on your excellent record of achievement so eloquently portrayed in the article itself. It is a saga of American business, and a great tribute to a fine gentleman, who by his high sense of service and business ethics is responsible for the measure of success achieved. I know that this record will serve as an inspiration to your two sons to carry on in the finest and best of traditions.

With every good personal wish for many years of good health, I am

Sincerely yours,

Peter I. Bukowski

In early 1955, Mayor Martin Kennelly was up for re-election again, and his office called on Best for his support. Kennelly wrote:

You have on several occasions indicated your interest in and approval of my administration. Many citizens, like yourself, have asked how they can help in this campaign in which I am a candidate for reelection. It has been suggested that if you would write to ten of your friends urging them to vote for me on February 22nd, who in turn would write or telephone ten of their friends, we would really have a force for good at work in the city.

On February 7, Best answered the mayor and enclosed a check. Two days later Kennelly wrote him:

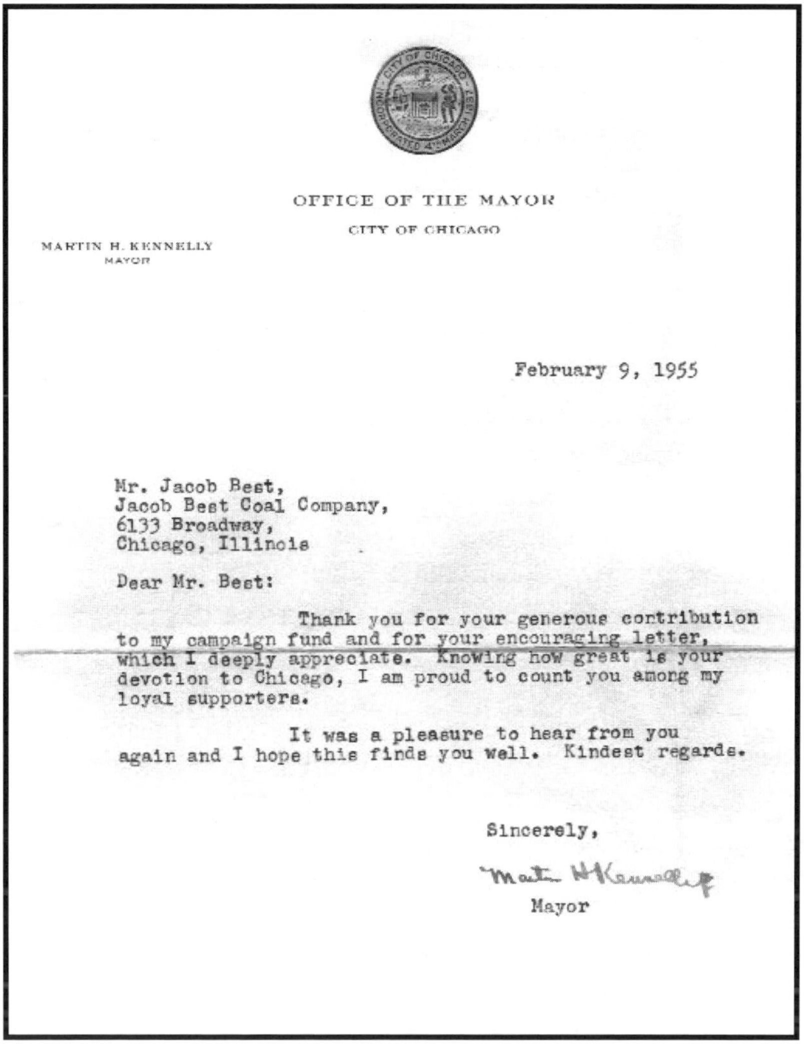

OFFICE OF THE MAYOR

CITY OF CHICAGO

MARTIN H. KENNELLY
MAYOR

February 9, 1955

Mr. Jacob Best,
Jacob Best Coal Company,
6133 Broadway,
Chicago, Illinois

Dear Mr. Best:

Thank you for your generous contribution to my campaign fund and for your encouraging letter, which I deeply appreciate. Knowing how great is your devotion to Chicago, I am proud to count you among my loyal supporters.

It was a pleasure to hear from you again and I hope this finds you well. Kindest regards.

Sincerely,

Martin H Kennelly
Mayor

Mayor Kennelly letter.

In the end, Richard J. Daley would deny Kennelly a third term. Such was the world of politics. Jacob suffered a more personal loss on Christmas night, 1955.

Jacob Best Coal Company celebrating July 4th at the Edgewater Yard. Like many other second-generation Americans, Best felt a ready patriotism.

The newly renovated sanctuary of St. Pauls, dating to 1898, caught fire and suffered extensive damage. Jacob, of course, felt obligated to help with the rebuilding campaign. There were three words inscribed in German on a bell that crashed down from the tower: *Glaube, Hoffnung, and Liebe*. Faith Hope and Love were the very qualities needed for the congregation to recover from the second burning of their church.

With his church efforts, Jacob began to curtail his civic life. On May 28, 1956, he informed Mayor Richard J. Daley that, due to poor health, he believed it desirable to retire from the board of the Chicago Public Library. It was, he added, a privilege to serve Chicago this way. The Mayor responded:

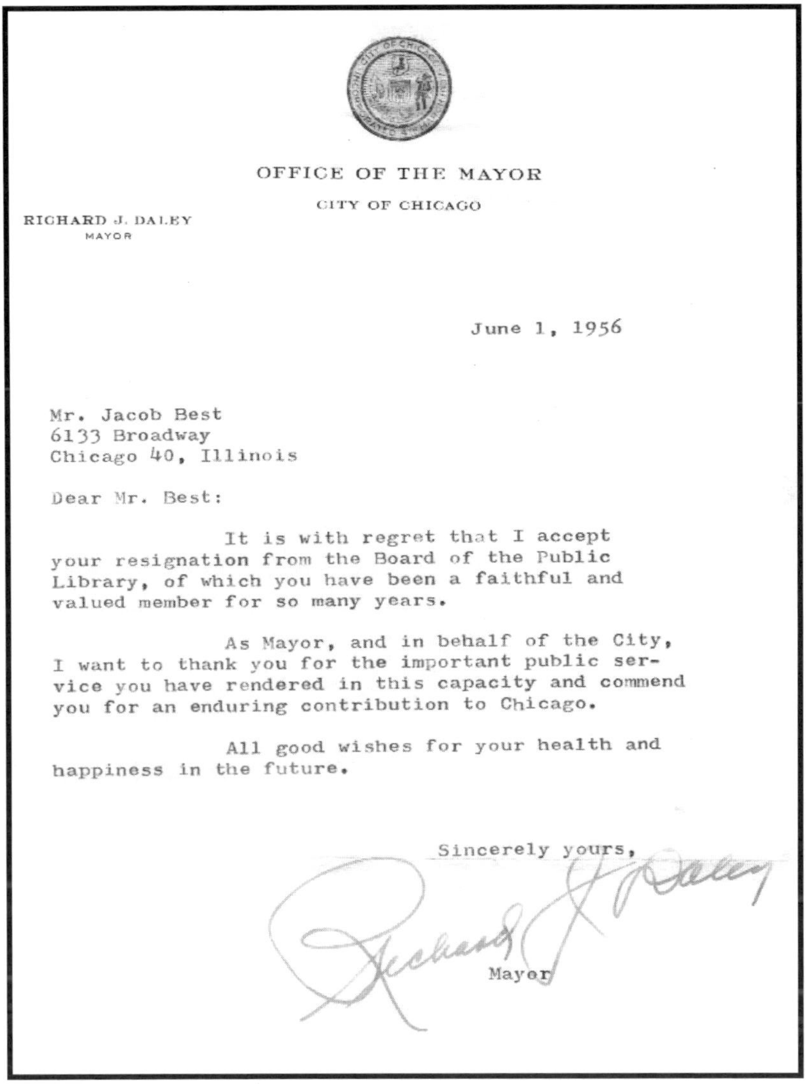
OFFICE OF THE MAYOR

CITY OF CHICAGO

RICHARD J. DALEY
MAYOR

June 1, 1956

Mr. Jacob Best
6133 Broadway
Chicago 40, Illinois

Dear Mr. Best:

It is with regret that I accept your resignation from the Board of the Public Library, of which you have been a faithful and valued member for so many years.

As Mayor, and in behalf of the City, I want to thank you for the important public service you have rendered in this capacity and commend you for an enduring contribution to Chicago.

All good wishes for your health and happiness in the future.

Sincerely yours,

Mayor

Mayor Daley Letter.

On January 24, 1957, Best attended a ceremony for the dedication of the Harold A. Bezazian Branch Library at 1226 West Ainslie Street; Jacob had served as Chairman of the Committee on

Buildings and Grounds. This was the last of his public appearances in the service of the city. Best now lived in his Evanston apartment, attended family affairs and lived quietly. Jacob had always said that he wanted to live long enough to see Best Coal celebrate fifty years and his only grand daughter get married. In that he would not be disappointed.

A dream come true: The marriage of Powell Bridges (from left) to granddaughter Barbara Ann, February 27, 1960. Flanking Jacob Best to his right are Carol Best and her husband Thilo Howard Best, Jacob's grandson.

Jacob Best of Chicago:
Businessman, Philanthropist, Native Son.

BEST AND HIS CHICAGO LEGACY

Jacob Best was an entrepreneur of character and sound business ethics. This pipe-smoking, unaffecting "go-getter" radiated efficiency and even invincibility. He was a man of few words and a believer in cooperative work. If the coal business ever needed a white knight, it found the perfect candidate in Jacob Best.

Perhaps the highest praise to give Jacob would be to say that during the 1920s, the age of Teapot Dome and Wall Street excess, he did things differently. *The Black Diamond* reported:

> *Here is the true ring of an honest effort to conduct a coal enterprise along lines that unmistakably exhibit the Golden Rule in business said a leading citizen of Rogers Park, as he handed me [the author of the article] a letter. The letter read as follows:*
>
> *'On January 21st we delivered to you two tons of Pocahontas Mine Run Coal. We mean to give our customers satisfaction, to follow our fuel beyond the mere sale of it, with a positive assurance that it shall be and continue to be satisfactory. We try to carry only such coal as will give satisfaction, to our customers, and if any delivery proves to be unsatisfactory to you such coal is unworthy of space in our yard. If you will let us know if your delivery was not carefully made or of any faults our coal may have, it will aid us in future selection and will be considered favor.'*

What held true in the 1920s continued for another forty years. If Jacob learned one lesson from Marshall Field, it was that the customer is always right. It was a philosophy fit for coal as well as fur.

Best humbly gave credit to all the men around him who lent their skills and talents to the building of one of Chicago's, as well as America's, leading retail/wholesale coal operations. All three children-Thilo, Jacob and Bernice-had equal shares in the ownership of the business. Thilo and Jake were fortunate to hold jobs through the Great Depression, and they happily shoveled coal from dawn to dusk.

The Black Diamond wrote, "In spite of the fact that the combined Edgewater and Mayfair operations grew somewhat beyond Jacob Best's most optimistic hopes and perhaps a future that even F.S. Peabody didn't foresee for the firm 45 years ago, the Best Company had never let its size or success get out of hand. The five-ton customer wasn't dropped to the bottom of the delivery list when the fifty ton orders started rolling in, and the furnace with the 22-inch retort was inspected just as carefully as the 75 HP boiler."

As a philanthropist, Jacob Best gave back to the hometown that bought his coal. Jacob wanted to do right by his city and according to his faith. In 1924 *National Magazine* noted, "Best is always found back of every civic advancement, and has taken an active part in the building of the community. He is now president of the Kiwanis Club of Rogers Park. He is particularly interested in the Boy Scouts of America, and the last campaign for funds was chairman of the local drive, and succeeded in raising more money than any other division in Chicago." As a board member for the Chicago Public Library, Uhlich Boys Orphanage, St. Pauls House for the Elderly, and St. Pauls Church, Jacob gave of his time, talent and, when necessary, treasure.

Jacob Best (at center behind flag) with fellow Kiwanis.

In 1958 Jacob started a charitable foundation. To date, the Jacob Best Foundation of Chicago has given over $1.5 million to children's causes, specifically to orphanages in metropolitan Chicago. Many children's charities have benefited from the largess of Jacob Best, who understood the importance of a stable family life.

Jacob died May 3, 1963, in Evanston.

EPILOGUE

A year later, Thilo, Jake and Bernice Best decided to sell the company to the Lill Coal Company. It was a timely and shrewd business decision. Thilo and Jake sensed that the trend in energy was moving away from coal toward heating oil and natural gas. Ironically, many of the coal companies went bankrupt shortly after the Bests sold their fathers' company.

Perhaps the most enduring legacy of Jacob Best beyond his philanthropy is the way he made his company so much a part of the city fabric. A neighborhood newspaper offered this view of Best Coal on Broadway in the summer of 1955:

> *The Christmas and Halloween windows are truly a work of art. On appropriate days, a big beautiful American flag is stretched across the window. It's all the more impressive because so many merchants fail to display our flag on Memorial, 4th of July and other proper times. Right now, the Best Coal Company window sports a doll-size kitchen stove and a stoker hopper filled with Sahara Stoker Coal. Both items are part of a contest in which a lucky young lady under 12 will win the stove and a lucky coal user (maybe her Daddy) can win 5 ton of Sahara Coal by guessing how many pieces of coal are in the hopper. In order to be eligible to win the doll-size stove, the young lady must write 50*

words or less on why her Daddy prefers Sahara Coal. The best letter will win. And, hurry over to 6133 N. Broadway with Daddy or grandpa to see the unusual window. Enter the contest and you may be the winner.

It was the Best of times.

GENALOGY OF THE BEST FAMILY

Generation No. 1

1. JACOB BEST was born 1835 in Ramsan, Bavaria, Germany, and died Mar 23, 1887 in Chicago, Illinois-Lakeview District. He married (1) KATHERINE NESS Aug 19, 1869 in Hettenleidelheim, Germany. She was born 1844 in Hettenleidelheim, Bavaria, Germany, and died Jun 10, 1886 in Chicago, Illinois-Lakeview District. He married (2) MATHILDA LANGE Sep 7, 1886. She was born 1856.

Children of JACOB BEST and KATHERINE NESS are:

i. JACOB BEST, b. May 27, 1878, Chicago, Illinois; d May 3, 1963, Evanston, Illinois.

ii. TILLIE LUCY BEST, b. Nov 29, 1870, Chicago, Illinois; d. Oct 4, 1949, Chicago, Illinois.

iii. GEORGE BEST, b. May 23, 1872, Chicago, Illinois; d. Feb 8, 1943, Chicago, Illinois.

iv. MARYANNE "Mayme" BEST, b. 1880, Chicago, Illinois; d. 1944, Chicago, Illinois.

Generation No. 2

2. JACOB BEST was born May 27, 1878 in Chicago, Illinois, and died May 3, 1963 in Evanston, Illinois. He married HULDA HAMMERSTROM Jun 7, 1904 in Chicago, Illinois, daughter of WILLIAM HAMMERSTROM and MINNIE THILO. She was born Jan 7, 1880 in Chicago, Illinois, and died Apr 11, 1945 in Chicago, Illinois.

Children of JACOB BEST and HULDA HAMMERSTROM are:

- i. THILO MAX BEST, b. Nov 17, 1906, Chicago, Illinois; d. Sep 12, 1989, Chicago, Illinois.
- ii. JACOB HILMER BEST, b. Sep 15, 1909, Chicago, Illinois; d. May 6, 1995, Lake Forest, Illinois.
- iii. BERNICE BEST, b. Jul 31, 1914, Chicago, Illinois; d. Apr 16, 1974, Chicago, Illinois.

3. TILLIE LUCY BEST was born Nov 29, 1870 in Chicago, Illinois, and died Oct 4, 1949 in Chicago, Illinois. She married CHARLES E. SIMON Nov 16, 1886 in Chicago, Illinois. He was born Dec 26, 1862, and died Sep 21, 1909.

Children of TILLIE BEST and CHARLES SIMON are:

- i. LOUIS OLEDIA SIMON, infant death.
- ii. ODELIA SIMON, b. May 5, 1895; d. Dec 24, 1965.
- iii. LOLITA SIMON, b. Sep 14, 1898.
- iv. KATHRYN SIMON, b. Jan 8, 1904; d. Apr 3, 1968.

4. GEORGE BEST was born May 23, 1872 in Chicago, Illinois, and died Feb 8, 1943 in Chicago, Illinois. He married MARY GITZKE Jul 24, 1893. She was born Feb 26, 1870, and died May 2, 1935.

Children of GEORGE BEST and MARY GITZKE are:

i. GRACE BEST, b. Jul 7, 1894; d. Feb 14, 1954; m. ATTO FISCHER, Aug 13, 1913; b. Apr 13, 1881; d. May 4, 1950.

ii. RUTH BEST, b. Sep 27, 1898.

iii. LAURA BEST, b. Mar 8, 1902; d. Sep 23, 1951.

iv. GEORGE BEST, b. Apr 19, 1907; d. Oct 27, 1962; m. BILLIE ANN MOLSKY, Jan 8, 1936; b. Aug 18, 1911.

5. MARYANNE "Mayme" BEST was born 1880 in Chicago, Illinois, and died 1944 in Chicago, Illinois. She married MAX W. GOLTZ May 28, 1904. He was born Sep 18, 1873, and died Oct 21, 1955.

Child of MARYANNE "Mayme" BEST and MAX GOLTZ is:

i. MAX F. GOLTZ, b. Oct 31, 1907; d. Jul 2, 1947.

Generation No. 3

6. THILO MAX BEST was born Nov 17, 1906 in Chicago, Illinois, and died Sep 12, 1989 in Chicago, Illinois. He married BARBARA VIRJEAN CAMPBELL Oct 9, 1930 in Chicago, Illinois, daughter of BENJAMIN CAMPBELL and MABELLE SAWTELLE. She was born Jun 11, 1906 in Chicago, Illinois, and died Apr 3, 1991 in Chicago, Illinois.

Children of THILO BEST and BARBARA CAMPBELL are:

i. THILO HOWARD BEST, b. Aug 31, 1931, Chicago, Illinois.

ii. BARBARA ANN BEST, b. Mar 14, 1936, Evanston, Illinois.

7. JACOB HILMER BEST was born Sep 15, 1909 in Chicago, Illinois, and died May 6, 1995 in Lake Forest, Illinois. He married CLARA DOUGLAS CORNELL Jun 5, 1936 in Glencoe, Illinois. She was born May 19, 1911, and died Aug 12, 1987 in Lake Forest, Illinois.

Children of JACOB BEST and CLARA CORNELL are:

i. JERRY HILMER BEST, JR., b. Jul 21, 1937, Evanston, Illinois.

ii. SCOTT DOUGLAS BEST, b. Dec 19, 1941, Evanston, Illinois.

8. ODELIA SIMON was born May 5, 1895, and died Dec 24, 1965. She married JOHN HANSEN. He was born Sep 23, 1892, and died Feb 9, 1965.

Children of ODELIA SIMON and JOHN HANSEN are:

i. KENNETH HANSEN, b. May 4, 1921.

ii. RUSSELL HANSEN, b. Jul 20, 1925.

9. LOLITA SIMON was born Sep 14, 1898. She married WALTER EGGERT Aug 11, 1920. He was born Apr 10, 1893, and died May 12, 1967.

Child of LOLITA SIMON and WALTER EGGERT is:

i. CHARLES W. EGGERT, b. Apr 6, 1936.

10. KATHRYN SIMON was born Jan 8, 1904, and died Apr 3, 1968. She married EMORY KING CROCKER Sep 10, 1929. He was born Apr 28, 1903, and died Dec 26, 1966.

Children of KATHRYN SIMON and EMORY KING CROCKER are:

i. LORRAINE CROCKER, b. Mar 12, 1931.

ii. RICHARD CROCKER, b. Aug 26, 1938.

iii. NORMAN CROCKER, b. May 8, 1943.

11. RUTH BEST was born Sep 27, 1898. She married EINAR WALSTROM Nov 5, 1920. He was born Feb 11, 1895.

Child of RUTH BEST and EINAR WALSTROM is:

i. DONALD A. WALSTROM, b. Oct 2, 1927.

12. LAURA BEST was born Mar 8, 1902, and died Sep 23, 1951. She married ARTHUR M. WILBINS Dec 12, 1925. He was born May 23, 1901, and died Aug 31, 1970.

Child of LAURA BEST and ARTHUR WILBINS is:

i. BETTY B. HORTON WILBINS, b. Apr 3, 1908.

13. MAX F. GOLTZ was born Oct 31, 1907, and died Jul 2, 1947. He married ISOPHINE MAY RICHEY Jan 18, 1930 in Chicago, Illinois. She was born Aug 7, 1930 in Chicago, Illinois.

Child of MAX GOLTZ and ISOSPHINE RICHEY is:

i. VIRGINIA MAY GOLTZ, b. Mar 31, 1934, Chicago, Illinois.

Generation No. 4

14. BARBARA ANN BEST was born Mar 14, 1936 in Evanston, Illinois. She married SHELBY POWELL BRIDGES, Feb 27, 1960 in Chicago, Illinois, son of HENRY Percival BRIDGES and CORNELIA Shelby THOMAS. He was born Apr 29, 1932 in Johnson City, Tennessee.

Children of BARBARA BEST and SHELBY Powell BRIDGES are:

i. THOMAS SHELBY BRIDGES, b. Nov 4, 1960, Evanston, Illinois; m.
CYNTHIA TREADWELL, Jan 7, 1984, Indianapolis, Indiana; b. Nov 20, 1961, Evanston, Illinois.

ii. DAVID POWELL BRIDGES, b. Jan 15, 1963, Evanston, Illinois.

iii. ROBERT CAMPBELL BRIDGES, b. Jun 29, 1965, Evanston, Illinois; m.
DEBORAH SUZANNE KING, Sep 15, 1990, Evanston, Illinois; b. Dec 11, 1964, Wheatridge, Colorado.

iv. BARBARA CARYL BRIDGES, b. Feb 24, 1969, Evanston, Illinois; m.
JEROME DESSERICH, Sep 19, 1998, Evanston, Illinois; b. Dec 29, 1966, Cincinnati, Ohio.

15. THILO HOWARD BEST was born Aug 31, 1931 in Chicago, Illinois. He married (1) CAROL CHRISTIANSON. She was born Dec 12, 1937. He married (2) LOUISE STOUT, Dec 12, 1976. She was born Jul 16, 1929.

Children of THILO BEST and CAROL CHRISTIANSON are:

i. THILO DAVID BEST, b. May 14, 1961, Chicago, Illinois; m. BETSY ALEXANDER, Oct 29, 1988, Atlanta, Georgia; b. Jul 06, 1964, Atlanta, Georgia.

ii. DIANA LYNN BEST, b. Jul 26, 1964, Chattanooga, Tennessee; m. LEE AYERS, Aug 10, 1985, Chattanooga, Tennessee; b. Feb 12, 1957, Columbia, South Carolina.

16. JERRY HILMER BEST, JR. was born Jul 21, 1937 in Evanston, Illinois. He married JANET P. DONNELLY Jun 20, 1959 in Gross Point, Michigan. She was born Sep 13, 1937 in Mt. Clemens, Michigan. Children of JERRY BEST and JANET DONNELLY are:

i. JACOB HILMER BEST III, b. Mar 30, 1960, Evanston, Illinois; m. LORIE GOTH, Jan 7, 1984, Honolulu, Hawaii; b. Apr 10, 1961, Lincoln, Nebraska.

ii. PETER BENNETT BEST, b. Apr 5, 1962, Evanston, Illinois; m. DIANA GAYE RAMSBOT TOM, Aug 29, 1987, Macon, Georgia; b. May 10, 1962, Memphis, Tennessee.

iii. JULIE DONNELLY BEST, b. Mar 4, 1966, Ridgewood, New Jersey; m. DAN HUNT.

17. SCOTT DOUGLAS BEST was born Dec 19, 1941 in Evanston, Illinois. He married (1) GLORIA HITE Jun 19, 1964 in Detroit, Michigan. She was born Dec 6, 1941. He married (2) BARBARA BENNETT Jul 15, 1988 in Avalon, California (Catalina Island). She was born Feb 8, 1944.

Children of SCOTT BEST and GLORIA HITE are:

i. TODD THOMAS BEST, b. Nov 28, 1960, Royal Oak, Michigan; m. SUZANNE MCLAUGHLIN, Mar 14, 1987; b. Sep 6, 1960.

ii. BRYAN CORNELL BEST, b. May 27, 1966, Royal Oak, Michigan; m. KAREN FINKBEINER, Dec 28, 1991; b. Apr 6, 1967, Pittsburgh, Pennsylvania.

iii. DARRIN HOWARD BEST, b. Jan 1, 1969, Royal Oak, Michigan; m. RUTH STOCKDALE, Jul 8, 1995; b. Jul 23, 1969.

18. CHARLES W. EGGERT was born Apr 6, 1936. He married (LOIS) EILEEN SWARTZ Sep 28, 1963. She was born Oct 17, 1938, in Flint, Michigan.

Children of CHARLES EGGERT and EILEEN SWARTZ are:

i. SUSAN ELIZABETH EGGERT, b. Feb 10, 1965, Detroit, Michigan; m.

MARTIN DANIEL FEENEY, Mar 17, 1984; b. Mar 10, 1948, Boston, Massachusetts.

ii. KRISTIN EILEEN EGGERT, b. Apr 13, 1967, Warren, Michigan; m. BRADLEY GENE HAYS, May 18, 1996; b. Jan 6, 1957, Dekalb, Illinois.

iii. THOMAS CHARLES EGGERT, b. Jul 22, 1970, Cleveland, Ohio; m. TRACIE JACQUELINE LAVINN, Aug 13, 1994; b. May 6, 1970, St. Charles, Illinois.

19. Lorraine Jean Crocker was born Mar 12, 1931 in Los Angeles, California.

20. Richard Emory Crocker was born Aug 26, 1938 in Los Angeles, California. He married Gale Greenseid, Mar 19, 1977. She was born Oct 27, 1942, in Los Angeles, California.

Children of Richard Emory Crocker and Gale are:

i. HILARY CROCKER, b. Jul 3, 1980 in Los Angeles, California.

ii. JEFFERY CROCKER, b. Apr 15, 1982 in Los Angeles, California.

21. Norman King Crocker was born May 8, 1943. He married DIANE LIPPE born Sep 17, 1945.

Children of Norman King Crocker and DIANE LIPPE CROCKER are:

i. KATHERYN CROCKER, b. Sep 6, 1969 in Los Angeles, California.

ii. GLENN EMERY CROCKER, b. Jun 28, 1972 in Los Angeles, California.

iii. HEIDI CROCKER, b. May 28, 1974 in Los Angeles, California

22. DONALD A. WALSTROM was born Oct 2, 1927. He married JOY SMITH Oct 27, 1951. She was born Jul 8, 1928.

Children of DONALD WALSTROM and JOY SMITH are:

i. DONALD H. WALSTROM, b. Feb 8, 1954.

ii. DONNA J. WALSTROM, b. Jan 18, 1956.

iii. LAURA R. WALSTROM, b. Aug 4, 1961.

iv. DONALD A. WALSTROM, b. Aug 3, 1966.

23. BETTY B. HORTON WILBINS was born Apr 3, 1908.

Child of BETTY B HORTON WILBINS is:

i. RICHARD GEORGE WILBINS, b. Jun 6, 1928; m. EVELYN GEBBARD, Mar 24, 1950; b. Mar 21, 1925.

24. VIRGINIA MAY GOLTZ was born Mar 31, 1934 in Chicago, Illinois. She married JERRY W. LORIMOR Sep 7, 1963 in Ottawa, Iowa. He was born Feb 9, 1937 in Hedrick, Iowa.

Child of VIRGINIA GOLTZ and JERRY LORIMOR is:

i. SUSAN LOUISE LORIMOR, b. Apr 8, 1965, Ottawa, Iowa.

25. KENNETH HANSEN was born May 4, 1921. He married JEAN KATHERINE DUNBECK, born Dec 28, 1920, she died 2001.

Children of KENNTH HANSEN and GENE DUNBECK are:

i. KENNETH HANSEN

ii. KEVIN HANSEN

26. RUSSELL HANSEN was born Jul 20, 1925. He married CONNIE RICHMOND in 1948. She was born Apr 29, 1929.

Children of RUSSELL HANSEN and CONNIE are:

i. KAREN HANSEN, b. Aug 27, 1949; m. MICHAEL HOLDATT Nov 14, 1982; b. Sep 5, 1948.

ii. KRISTEN ANN HANSEN, b. Nov 14, 1953; m. KEN CLARENCE BINKLEY; Aug 8, 1970; b. Nov 27, 1950.

Generation No. 5

27. THOMAS SHELBY BRIDGES was born Nov 4, 1960 in Evanston, Illinois. He married CYNTHIA DENISE TREADWELL Jan 7, 1984 in Indianapolis, Indiana. She was born Nov 20, 1961 in Indianapolis, Indiana.

Children of THOMAS BRIDGES and CYNTHIA DENISE TREADWELL are:

i. SHELBY LEE BRIDGES, b. Aug 25, 1990, Evanston, Illinois.

ii. LAUREN MARIE BRIDGES, b. Jan 31, 1994, Evanston, Illinois.

28. ROBERT CAMPBELL BRIDGES was born Jun 29, 1965 in Evanston, Illinois. He married DEBORAH SUZANNE KING, Sep 15, 1990 in Evanston, Illinois. She was born Dec 11, 1964 in Wheatridge, Colorado.

Children of ROBERT BRIDGES and DEBORAH KING are:

i. EMILY GRACE BRIDGES, b. Feb 22, 1995, Evanston, Illinois.

ii. VICTORIA MORGAN BRIDGES, b. Jun 22, 1997, Evanston, Illinois.

29. BARBARA CARYL BRIDGES was born Feb 24, 1969 in Evanston, Illinois. She married JEROME DAVID DESSERICH Sep 19, 1998 in Evanston, Illinois. He was born Dec 29, 1966 in Cincinnati, Ohio.

Child of BARBARA CARYL BRIDGES and JEROME DAVID DESSERICH is:

i. ASHLEY MAE DESSERICH, b. Dec 3, 2000.

30. THILO DAVID BEST was born May 14, 1961 in Chicago, Illinois. He married BETSY ALEXANDER Oct 29, 1988 in Atlanta, Georgia. She was born Jul 6, 1964 in Atlanta, Georgia.

Children of THILO BEST and BETSY ALEXANDER are:

i. ELIZABETH STANFIELD BEST, b. Apr 16, 1993, Atlanta, Georgia.

ii. MEREDITH ASHBY BEST, b. May 3, 1996, Portland, Oregon.

31. DIANA LYNN BEST was born Jul 26, 1964 in Chattanooga, Tennessee. She married LEE AYERS Aug 10, 1985 in Chattanooga, Tennessee. He was born Feb 12, 1957 in Columbia, South Carolina.

Children of DIANA BEST and LEE AYERS are:

i. CLIFTON CAMPBELL AYERS, b. Apr 3, 1991.

ii. JACOB LEE AYERS, b. Mar 17, 1993.

iii. GREGORY GARRETT AYERS, b. May 6, 1996.

32. JACOB HILMER BEST III was born Mar 30, 1960 in Evanston, Illinois. He married LORIE GOTH Jan 7, 1984 in Honolulu, Hawaii. She was born Apr 10, 1961 in Lincoln, Nebraska.

Children of JACOB BEST and LORIE GOTH are:

i. BRANDT RYAN BEST, b. Jun 28, 1985, Torrance, California.

ii. COURTNEY M. BEST, b. Jun 3, 1988, Newport Beach, California.

iii. AUSTIN JACOB BEST, b. Dec 30, 1989, Poway, California.

33. PETER BENNETT BEST was born Apr 5, 1962 in Evanston, Illinois. He married DIANA GAYE RAMSBOTTOM Aug 29, 1987 in Macon, Georgia. She was born May 10, 1962 in Memphis, Tennessee.

Children of PETER BEST and DIANA RAMSBOTTOM are:

i. TUCKER BENNETT BEST, b. Jun 6, 1991, Santa Monica, California.

ii. SARA DIANA BEST, b. Jan 12, 1993, Santa Monica, California.

34. JULIE DONNELLY BEST was born Mar 4, 1966 in Ridgewood, New Jersey. She married DAN HUNT.

Children of JULIE BEST and DAN HUNT are:

i. TREVOR W. HUNT, b. Dec 23, 1992, Palm Springs, California.

ii. CARLY KATHLEEN HUNT, b. Aug 1, 1994, Palm Springs, California.

35. TODD THOMAS BEST, b. Nov 28, 1960, Royal Oak, Michigan; m. SUZANNE MCLAUGHLIN, Mar 14, 1987; b. Sep 6, 1960.

Children of TODD THOMAS BEST and SUZANNE MCLAUGHLIN are:

i. JACOB SCOTT BEST, Jul 29, 1993, Royal Oak, Michigan

ii. JOHN TODD BEST, Nov 20, 2001, Royal Oak, Michigan

36. BRYAN CORNELL BEST was born May 27, 1966 in Royal Oak, Michigan. He married KAREN FINKBEINER Dec 28, 1991. She was born Apr 6, 1967 in Pittsburgh, Pennsylvania.

Children of BRYAN BEST and KAREN FINKBEINER are:

i. BRANDON CORNELL BEST, b. Sep 2, 1995, Elk Grove Village, Illinois.

ii. BLAKE THOMAS BEST, b. Nov 25, 1997, Hershey, Pennsylvania.

iii. BRETT MAXWELL BEST, b. Apr 2, 2000, Royal Oak, Michigan.

37. DARRIN HOWARD BEST was born Jan 1, 1969 in Royal Oak, Michigan. He married RUTH STOCKDALE Jul 8, 1995. She was born Jul 23, 1969.

Children of DARRIN BEST and RUTH STOCKDALE are:

i. RACHEL MADISON BEST, b. Sep 18, 1997, Royal Oak, Michigan.

ii. ANDREW TRAVIS BEST, b. May 6, 2000, Royal Oak, Michigan.

38. RICHARD GEORGE WILBINS was born Jun 6, 1928. He married EVELYN GEBBARD Mar 24, 1950. She was born Mar 21, 1925.

Children of RICHARD WILBINS and EVELYN GEBBARD are:

i. RICHARD GEORGE WILBINS, JR., b. Sep 25, 1951.

ii. MICHAEL KIRK WILBINS, b. Jan 6, 1953.

iii. PEGGY ANN WILBINS, b. Feb 10, 1954.

v. VICKIE LYNN WILBINS, b. Jul 1, 1960.

39. MARTIN DANIEL FEENEY was born Mar 10, 1948. He married SUSAN ELIZABETH Mar 17, 1984. She was born Feb 10, 1965.

Children of MARTIN DANIEL FEENEY and EVELYN GEBBARD are:

i. ROBERT CHARLES THOMAS FEENEY, b. Aug 9, 1985, Wheaton, Illinois.

ii. CAROLINE EILEEN FEENEY, b. May 25, 1990, Des Moines, Iowa.

40. BRADLEY GENE HAYS was born Jan 6, 1957. He married KRISTEN EILEEN May 18, 1996. She was born Apr 13, 1967.

Children of BRADLEY GENE HAYS and KRISTEN EILEEN are:

i. MADELINE KATHERINE, b. Nov 27, 1998, Winfield, Illinois.

41. MICHAEL A. JACKSON was born in 1969, Los Angeles, California. He married KATHERYN CROCKER, Oct 17, 1998. She was born Sep 6, 1969 in Los Angeles, California.

Children of MICHAEL A. JACKSON and KATHERYN CROCKER JACKSON are:

i. MAX ALEXANDER JACKSON, b. Aug. 18, 2000, Los Angeles, California.

42. KAREN HANSEN, was born Aug 27, 1949; m. MICHAEL HOLDATT Nov 14, 1982; He was born Sep 5, 1948.

Children of KAREN HANSEN and MICHAEL HOLDATT are:

i. MICHELL TERESA HOLDATT, b. Dec 23, 1983.

ii. KRISTINE ANN HOLDAT, b. AUG 30, 1985.

iii. COREY HOLDATT, b. Dec 17, 1987.

43. KRISTEN ANN HANSEN was born Nov 14, 1953; She married KEN CLARENCE BINKLEY Aug 8, 1970. He was born Nov 27, 1950.

Children of KRISTEN ANN HANSEN and KEN CLARENCE BINKLEY are:

i. DEREK RUSSELL BINKLEY, b. Sep 3, 1983.

ii. MEREDITH ANN BINKLEY, b. Sep 27, 1985.

iii. MEGAN BINKLEY, Nov 14, 1989.

ENDNOTES

Chapter 1

"Jacob Best and his": Heiraten Marriage Book, Hettenleidelheim Kirchenbuch, #0489427. 1869.

"People talked of": Perry Duis, *The Saloon* (1999), 152.

"The experience of": *Hartmut Keil and John B. Jentz,* German Workers in Chicago: A Documentary History of Working-Class Culture from 1850 to World War I (1988), 21.

"This Indenture Witnesseth": Cook County Recorder of Deeds. Book 1481, page 209, no. 559669, 1884.

"Lots Twenty Five": ibid, Book 1603, Page 9, no. 650474, 1885.

"The variety": Melvin G. Holli and Peter d'A Jones, *Ethnic Chicago: A Multicultural Portrait* (1995), 505.

"The fastest way": Duis, *Saloon*, 141.

"A Large Stock": Chicago Directory 1859, 48.

"the wagons go": Duis, *Saloon*, 67.

"For whole families": ibid, 120.

"This Society shall": *Constitution and By-Laws of the Saloon Keepers, 4*

"respectable citizens": ibid, 4.

"The children performed": Duis, *Saloon*, 49.

"If your regular": ibid, 58.

"The story goes": Holli and Jones, *Ethnic Chicago*, 505.

"men ravenously hungry": F.L. Kalvelage, *The Annals of St. Boniface Parish 1862-1926*, 6.

"which rang from": ibid, 10.

"had tables and": Duis, *Saloon*,154-155.

"St. Boniface": Kalvelage, *Annals*, 43.

"December 28, 1888": Cook County Recorder of Deeds. Book 2425, page 383, no. 1045062, 1888.

Chapter 2

"People began calling": Donald L. Miller, *City of the Century: The Epic of Chicago and the Making of America* (1996), 188.

"while people hurried": ibid, 185.

"Leaving work": ibid, 341.

"were a diverse": *Duis, Challenging Chicago*, 287-288.

"full share": Miller, *City of the Century*, 163.

"President Grover Cleveland": ibid, 490.

"Generally plain": *Official Guide To The World's Columbian Exposition* (1893), 150.

"The Krupp gun": ibid, 119.

"Always keep at": Miller, *City of the Century*, 222.

"To Whom It May Concern": Jacob Best Collection, Group 4 (hereinafter JBC/GR), 125.

"To the average Easterner": Marshall Field and Company advertisement in author's possession (1900).

"I here send": JBC/GR.1, 46-48.

"J. Best": JBC/GR 1, 11-14.

"Dear Friend Hulda": JBC/GR 1, 126-128.

"No you did not": JBC/GR 3, 109-110.

"The latest talk": JBC/GR 4, 29.

"We have had": JBC/GR 4, 59-60.

"Papa formed a trust": JBC/GR 4, 103-105.
"the lobster died": Duis, *Challenging Chicago*, 146.
"Enclosed you will": JBC/GR 1, 77-78.
"Vater schaft nicht": JBC/GR 1, 79.
"Maryanne begs": JBC/GR 3, 113.
"Words could not": JBC/GR 1, 28-30.
"It has been": Philip Henrici Restaurant, 25th Anniversary promotional (1894).
"It will be": JBC/GR 1, 103.
"Indeed I did": JBC/GR 1, 130.
"Answering yours": JBC/GR 3, 6-7.
"But Hulda": JBC/GR 3, 11.
"I have so": JBC/GR 3, 22-23.
"Was sincerely pleased": JBC/GR 3, 25-26.
"I was very sorry": JBC/GR 3, 32-34.
"Have referred Miss": JBC/GR 3, 39.
"However, you know": JBC/GR 3, 46.
"I am pleased": JBC/GR 3, 103-104.
"Oh how nice": JBC/GR 3, 106.
"Happy, well I dare": JBC/GR 3, 94.
"Do you know": JBC/GR 1, 41-43.
"you know I love": JBC/GR 3, 81-83.
"My dear Hulda": JBC/GR 4, 9-16.
"As early as 1834": Duis, *Challenging Chicago*, 226.
"most assuredly": JBC/GR 1, 16-18.
"Enclosed you will": JBC/GR 1, 5-6.
"Special Entertainments": JBC/GR 1, 7.
"The last letter": JBC/GR 1, 24-26.

"Maryanne and I": JBC/GR 1, 81-83.
"But as it": JBC/GR 1, 20-22.
"How did you feel": JBC/GR 1, 85-86.
"Your letter of": JBC/GR 1, 90
"Had a good": JBC/GR 3, 115.
"THE PRESENCE": JBC/GR 4, 24.
"I am pleased": JBC/GR 1, 137-138.
"Do you realize": JBC/GR 3, 77-79.
"But sweet one": JBC/GR 1, 32-39.
"ST. PAULS CHURCH": JBC/GR 4, 33.
"St. Pauls Church League": JBC/GR 4, 112.
"I have received": JBC/GR 4, 75.
"Your Mama": JBC/GR 4, 38-39.
"Father's Hand": JBC/GR 1, 11-14.
"I am very": JBC/GR 1, 51-53.
"Five weeks have passed": JBC/GR 1, 94-95.
"In reference": JBC/GR 1, 32-39.
"when labor": "A Coal Concern," *The Black Diamond*, 418.

Chapter 3

"America's and Chicago's": Miller, *City of the Century*, 243.
"Jacob Best, At": Muriel Herzberg, "Pioneer North Side Coal Firm Gets Modern Look: Jacob Best Co. Has Face-Lifting, Recalls Opening 45 Years Ago" in *Edgewater Uptown News* (September 14, 1954, v. 52, n. 38).
"delivered at": "A Coal Company That Stays Clean. 431 South Dearborn Street, Chicago, Illinois" in *The Black Diamond* (January 1929).

"I was impressed": ibid.

"nothing ornate": ibid.

"Like a violent": Melvin G. Holli and Peter d'A Jones, *Ethnic Chicago* (Revised and Expanded, 1984), 501.

"American patriotism": R. Thomas Henry, *Making a Joyful Sound in the City: The History of St. Pauls United Church of Christ* (1996), 20.

"rallied behind": ibid, 48.

"With new apartments": "Best Company Lives Up To Name For 45 Years. 431 South Dearborn Street, Chicago, Illinois" in *The Black Diamond* (August 1954), 22-26.

"In conjunction with": *The Howard, The Official Howard District Business Association A Weekly Newspaper*, August 4, 1927.

"Best on one": "Best Company Lives Up To," *The Black Diamond,* 22-26.

"Dear Friend: All": Jacob Best Scrapbook (hereinafter JBS), Best Coal Advertisement, September 13, 1927.

"During the nineteen years": "A Coal Company That Stays Clean."

"Next Monday morning": "May 25, 1929, *The Black Diamond* Retailers Will Convene Next Monday" in *The Black Diamond* (v. 28, n. 22).

"If we are": "The Black Diamond National Retail Meet Is History Maker, June 1, 1929" in *The Black Diamond* (v. 28, n. 22).

"a campaign such": ibid.

"rare executive ability": inscription on platter, in possession of author's family.

"My dear friend": JBS, Rudolph A. John, June 10, 1929.

"For more than": JBS, ad for JACOB BEST COAL Co., May 12,

1933.

"Jacob Best, president": "Truck Sign Helps Sell 'Best Coal'" in *The Black Diamond* (October 8, 1938, n. 8), 21.

"My Dear Friend": JBS, Rudolph A. John, December 10, 1935.

"Distinguished Service Cross": Henry, *Making a Joyful Sound*, 31.

"Office of": JBS, Office of the Mayor, City of Chicago/ Edward J. Kelly, June 23, 1943.

"Dear Bernice: Uncle": JBS, Macatawa, Michigan, June 27, 1945.

"Dear Bernice: You": ibid, July 11, 1945.

"Dear Bernice: It": ibid, July 25, 1945.

"Office of": JBS, Office of the Mayor, City of Chicago/ Martin H. Kennelly, June 15, 1950.

"Dear Mr. Best": JBS, St. Pauls Church, July 19, 1952.

"Dear Jake": JBS, Charles M. Moderwell, August 16, 1954.

"Dear Jack": JBS, Peter I. Bukowski, September 7, 1954.

"You have on": JBS, Mayor Martin Kennelly, January 17, 1955.

"Dear Mr. Best": JBS, Office of the Mayor, City of Chicago/ Martin H. Kennelly, February 9, 1955.

"June 1, 1956": JBS, Office of the Mayor, City of Chicago/ Richard J. Daley, June 1, 1956.

"Here is the": "A Coal Concern," 418-425.

"In spite of": "A Coal Company That Stays Clean," *The Black Diamond,* 22-26.

"Best is always": "A Coal Concern,"*The Black Diamond,* 425.

"The Christmas and Halloween": *Gossip Sheet, This 'N That* of the Granville-Broadway Business Association, June 14, 1955.

INDEX

ABOUT THE AUTHOR

David P. Bridges was raised in the suburb of Wilmette, north of Chicago. He has a BS in Agricultural Economics from the University of Kentucky, Lexington, Kentucky and a Masters in Divinity from Louisville Presbyterian Theological Seminary, Louisville, Kentucky. He studied at the University of Chicago Divinity School and has spent fifteen years preaching and teaching the Gospel as an ordained Presbyterian minister.

The author is an avid outdoorsman, hunter and fisherman. Mr. Bridges is single and resides in Alexandria, Virginia, and is a minister at Westminster Presbyterian Church. He has completed a second family history: *The Bridges of Washington County, Spanning Work and Nature.* He is currently writing: *Major James Breathed, Civil War Hero of General J.E.B. Stuart's 1st Virginia Cavalry & 1st Horse Artillery, C.S.A.*

David P. Bridges
DPBridges@AOL.com